T0335581

Ordinary Differential Equations and Boundary Value Problems

Volume I: Advanced Ordinary Differential Equations

TRENDS IN ABSTRACT AND APPLIED ANALYSIS

ISSN: 2424-8746

Series Editor: John R. Graef
The University of Tennessee at Chattanooga, USA

This series will provide state of the art results and applications on current topics in the broad area of Mathematical Analysis. Of a more focused nature than what is usually found in standard textbooks, these volumes will provide researchers and graduate students a path to the research frontiers in an easily accessible manner. In addition to being useful for individual study, they will also be appropriate for use in graduate and advanced undergraduate courses and research seminars. The volumes in this series will not only be of interest to mathematicians but also to scientists in other areas. For more information, please go to http://www.worldscientific.com/series/taaa

Published

Vol. 7   *Ordinary Differential Equations and Boundary Value Problems*
          Volume I: Advanced Ordinary Differential Equations
          by John R. Graef, Johnny Henderson, Lingju Kong &
          Xueyan Sherry Liu

Vol. 6   *The Strong Nonlinear Limit-Point/Limit-Circle Problem*
          by Miroslav Bartušek & John R. Graef

Vol. 5   *Higher Order Boundary Value Problems on Unbounded Domains:*
          Types of Solutions, Functional Problems and Applications
          by Feliz Manuel Minhós & Hugo Carrasco

Vol. 4   *Quantum Calculus:*
          New Concepts, Impulsive IVPs and BVPs, Inequalities
          by Bashir Ahmad, Sotiris Ntouyas & Jessada Tariboon

Vol. 3   *Solutions of Nonlinear Differential Equations:*
          Existence Results via the Variational Approach
          by Lin Li & Shu-Zhi Song

Vol. 2   *Nonlinear Interpolation and Boundary Value Problems*
          by Paul W. Eloe & Johnny Henderson

Vol. 1   *Multiple Solutions of Boundary Value Problems:*
          A Variational Approach
          by John R. Graef & Lingju Kong

Trends in Abstract
and Applied Analysis
Volume **7**

Ordinary Differential Equations and Boundary Value Problems

Volume I: Advanced Ordinary Differential Equations

John R Graef

University of Tennessee at Chattanooga, USA

Johnny Henderson

Baylor University, USA

Lingju Kong

University of Tennessee at Chattanooga, USA

Xueyan Sherry Liu

St Jude Children's Research Hospital, USA

World Scientific

NEW JERSEY · LONDON · SINGAPORE · BEIJING · SHANGHAI · HONG KONG · TAIPEI · CHENNAI · TOKYO

Published by

World Scientific Publishing Co. Pte. Ltd.
5 Toh Tuck Link, Singapore 596224
USA office: 27 Warren Street, Suite 401-402, Hackensack, NJ 07601
UK office: 57 Shelton Street, Covent Garden, London WC2H 9HE

Library of Congress Cataloging-in-Publication Data
Names: Graef, John R., 1942–   author.
Title: Ordinary differential equations and boundary value problems / by John R. Graef
    (University of Tennessee at Chattanooga, USA) [and three others].
Description: New Jersey : World Scientific, 2018– | Series: Trends in abstract and
    applied analysis ; volume 7 | Includes bibliographical references and index.
    Contents: volume 1. Advanced ordinary differential equations
Identifiers: LCCN 2017060286 | ISBN 9789813236455 (hc : alk. paper : v. 1)
Subjects: LCSH: Differential equations. | Boundary value problems.
Classification: LCC QA372 .O7218 2018 | DDC 515/.352--dc23
LC record available at https://lccn.loc.gov/2017060286

British Library Cataloguing-in-Publication Data
A catalogue record for this book is available from the British Library.

For any available supplementary material, please visit
http://www.worldscientific.com/worldscibooks/10.1142/10888#t=suppl

Desk Editors: V. Vishnu Mohan/Kwong Lai Fun

Typeset by Stallion Press
Email: enquiries@stallionpress.com

Printed in Singapore

Dedication

John Graef dedicates this work to his wife Frances and his Ph.D. advisor T. A. Burton.

Johnny Henderson's dedication first is to his friend, Allan C. Peterson, and second is to his cadre of doctoral students: Jeffrey Allen Ehme, Anjali Datta, Eric Roger Kaufmann, Kuo-Chuan William Yin, Rena Denis Taunton Reid, Feng-Chun Charlie Fang, Tuwaner Mae Hudson Lamar, Susan Denese Stenger Lauer, John Marcus Davis, Alvina M. Johnson Atkinson, Nickolai Kosmatov, Kathleen Marie Goeden Fick, Ana Maria Maturana Tameru, Parmjeet Kaur Singh Cobb, Basant Kumar Karna, Ding Ma, Michael Jeffery Gray, Mariette R. Maroun, John Emery Ehrke, Curtis John Kunkel, Britney Jill Hopkins, Jeffrey Wayne Lyons, Jeffrey Thomas Neugebauer, Xueyan Sherry Liu, Shawn Michael Sutherland, Charles Franklin Nelms, Jr. and Brian Christopher Pennington.

Lingju Kong dedicates this book to those mathematicians who influenced his research.

Xueyan Liu's dedication is to her Ph.D. advisor Johnny Henderson, MS advisor Binggen Zhang, postdoctoral supervisor Hui Zhang, colleague Deo Kumar Srivastava, and former colleagues John Graef, Lingju Kong, Min Wang, Cuilan Gao, Jin Wang, Andrew Ledoan, Xuhua Liu, academic friends Richard Avery, Douglas Anderson, Yu Tian, and academic cousins Jeffrey Thomas Neugebauer, Jeffrey Wayne Lyons, and Shawn Michael Sutherland.

Preface

In this work we give a treatment of the theory of ordinary differential equations (ODEs) that is appropriate to use for a first course at the graduate level as well as for individual study. Written in a somewhat chatty way, we hope the reader will find it to be a captivating introduction to this fascinating area of study. A number of nonroutine exercises are dispersed throughout the book to help the reader explore and expand his knowledge.

We begin with a study of initial value problems for systems of differential equations including the Picard and Peano existence theorems. The continuability of solutions, their continuous dependence on initial conditions, and their continuous dependence with respect to parameters are presented in detail. This is followed by a chapter on the differentiability of solutions with respect to initial conditions and with respect to parameters. Comparison results and differential inequalities come next.

Linear systems of differential equations are treated in detail as is appropriate for a study of ODEs at this level. We believe that just the right amount of basic properties of matrices is introduced to facilitate the study matrix systems and especially those with constant coefficients. Floquet theory for linear periodic systems is presented and used to study nonhomogeneous linear systems.

Stability theory of first order and vector linear systems are considered. The relationships between stability of solutions, uniform stability, asymptotic stability, uniformly asymptotic stability, and strong stability are discussed and illustrated with examples. A section on the stability of vector linear systems is included. The book concludes with a chapter on perturbed systems of ODEs.

A second volume devoted to boundary value problems is to follow. It can be used as a "stand alone" work or as a natural sequel to what is presented here.

John R. Graef
Johnny Henderson
Lingju Kong
Xueyan "Sherry" Liu

Contents

Preface   vii

1. Systems of Differential Equations   1

    1.1   Introduction 1

    1.2   The Initial Value Problem (IVP) 2

    1.3   The Picard Existence Theorem 4

    1.4   The Peano Existence Theorem 18

2. Continuation of Solutions and Maximal Intervals
of Existence   29

    2.1   Continuation of Solutions 29

    2.2   Kamke Convergence Theorem 37

    2.3   Continuous Dependence of Solutions on Initial
Conditions 42

    2.4   Continuity of Solutions wrt Parameters 45

3. Smooth Dependence on Initial Conditions and Smooth
Dependence on Parameters   51

    3.1   Differentiation of Solutions wrt Initial Conditions 51

    3.2   Differentiation of Solutions wrt Parameters 55

    3.3   Maximal Solutions and Minimal Solutions 61

4. Some Comparison Theorems and Differential Inequalities   65

    4.1   Comparison Theorems and Differential Inequalities 65

    4.2   Kamke Uniqueness Theorem 73

5. Linear Systems of Differential Equations   77

    5.1   Linear Systems of Differential Equations 77

5.2   Some Properties of Matrices   82

5.3   Infinite Series of Matrices and Matrix-Valued
      Functions   85

5.4   Linear Matrix System   88

5.5   Higher Order Differential Equations   96

5.6   Systems of Equations with Constant Coefficient
      Matrices   103

5.7   The Logarithm of a Matrix   116

6.   Periodic Linear Systems and Floquet Theory   121

6.1   Periodic Homogeneous Linear Systems
      and Floquet Theory   121

6.2   Periodic Nonhomogeneous Linear Systems
      and Floquet Theory   126

7.   Stability Theory   135

7.1   Stability of First Order Systems   135

7.2   Stability of Vector Linear Systems   140

8.   Perturbed Systems and More on Existence
     of Periodic Solutions   149

8.1   Perturbed Linear Systems   149

Bibliography   163

Index   165

Chapter 1

Systems of Differential Equations

1.1 Introduction

We shall be concerned with solutions of systems of ordinary differential equations (ODE's). Let $f(t,x)$ be defined on some set $D \subseteq \mathbb{R} \times \mathbb{R}^n$, in the sense that $t \in \mathbb{R}$ and $x \in \mathbb{R}^n$ (i.e., x is an n-dimensional vector). We shall record x as a column vector, so

$$x = \begin{bmatrix} x_1 \\ x_2 \\ \vdots \\ x_n \end{bmatrix}.$$

Moreover, let $f : D \to \mathbb{R}^n$, so that f is also a column vector, and in particular,

$$f(t,x) = \begin{bmatrix} f_1(t,x) \\ f_2(t,x) \\ \vdots \\ f_n(t,x) \end{bmatrix} = \begin{bmatrix} f_1(t,x_1,\ldots,x_n) \\ f_2(t,x_1,\ldots,x_n) \\ \vdots \\ f_n(t,x_1,\ldots,x_n) \end{bmatrix}.$$

Definition 1.1 (Classical Solution). An n-dimensional vector-valued function, $\varphi(t)$, is said to be *a solution of the differential equation (D.E.)* $x' = f(t,x)$ on an interval I in case:

(1) $\varphi \in C^{(1)}(I)$,
(2) $(t, \varphi(t)) \in D$, for all $t \in I$,
(3) $\varphi'(t) \equiv f(t, \varphi(t))$ on I.

Now in the above definition,

$$x' = f(t,x) \iff \begin{cases} \dfrac{dx_1}{dt} = f_1(t, x_1, x_2, \ldots, x_n), \\[2mm] \dfrac{dx_2}{dt} = f_2(t, x_1, x_2, \ldots, x_n), \\[1mm] \qquad\vdots \\[1mm] \dfrac{dx_n}{dt} = f_n(t, x_1, x_2, \ldots, x_n). \end{cases}$$

The solution described in the definition above is called *a classical solution.* Although we will not be concerned with such, we can discuss a solution in a more general sense (*the almost everywhere* (*a.e.*) or *Lebesgue sense*). In that case, such a solution φ of $x' = f(t,x)$ would satisfy:

(1) φ is absolutely continuous on I,

(2) $(t, \varphi(t)) \in D$, for all $t \in I$,

(3) $\varphi'(t) = f(t, \varphi(t))$ a.e. on I.

Exercise 1. Let $f(t, x)$ be continuous on $D \subseteq \mathbb{R} \times \mathbb{R}^n$ and let $\varphi(t)$ be an a.e. solution on an interval I. Prove $\varphi(t)$ is a classical solution.

We now shall begin our work towards the local existence theorems.

1.2   The Initial Value Problem (IVP)

Given $f(t, x)$ defined on $D \subseteq \mathbb{R} \times \mathbb{R}^n$ and $(t_0, x_0) \in D$, we seek an interval I and an n-vector function $\varphi(t)$ such that $t_0 \in I$, $\varphi(t_0) = x_0$, $\varphi \in C^{(1)}(I)$, $(t, \varphi(t)) \in D$ on I, and $\varphi'(t) \equiv f(t, \varphi(t))$ on I. Such a vector-valued function $\varphi(t)$ is said to be *a solution of the initial value problem* (*IVP*) *on I* and we usually indicate this IVP by

$$\begin{cases} x' = f(t, x), \\ x(t_0) = x_0. \end{cases} \tag{1.1}$$

Consider now the relationship between a solution of IVP (1.1) and a solution of an integral equation. Assume that $f(t, x)$ is continuous on D and that $\varphi(t)$ is a solution of IVP (1.1) on the interval I. Then, for all $t \in I$, we have $\varphi'(s) \equiv f(s, \varphi(s))$ on the interval with endpoints t_0 and t. By the continuity of $f(s, \varphi(s))$, we have that $\varphi'(s)$ is integrable over the interval with endpoints t_0 and t, and

$$\int_{t_0}^{t} \varphi'(s)\, ds = \int_{t_0}^{t} f(s, \varphi(s))\, ds,$$

which yields

$$\varphi(t) - \varphi(t_0) = \int_{t_0}^t f(s, \varphi(s))\, ds,$$

that is,

$$\varphi(t) = x_0 + \int_{t_0}^t f(s, \varphi(s))\, ds, \quad \text{for all } t \in I. \tag{1.2}$$

In particular, if $\varphi \in C^{(1)}(I)$ and is also a solution of IVP (1.1) on I, then $\varphi \in C(I)$ and is a solution of the integral equation (1.2) on I.

Conversely, if $\varphi \in C(I)$, $\text{Graph}(t, \varphi(t)) \subseteq D$, and φ satisfies (1.2) on I, then $\varphi(t_0) = x_0$; also, since $f \in C(D)$, we have that $f(t, \varphi(t)) \in C(I)$, and consequently $\int_{t_0}^t f(s, \varphi(s))\, ds$ is continuously differentiable. Thus,

$$\varphi \in C^{(1)}(I) \quad \text{and} \quad \varphi'(t) = \frac{d}{dt}\left(x_0 + \int_{t_0}^t f(s, \varphi(s))\, ds\right) = f(t, \varphi(t)).$$

Thus, φ is a solution of IVP (1.1).

Before discussing the local existence theorems, we will briefly consider some properties of "norms".

Definition 1.2. Let V be a vector space. Then *a norm* $\|\cdot\| : V \to \mathbb{R}$ has the properties:

(1) $\|v\| \geq 0$, and $\|v\| = 0$ if and only if $v = 0$,
(2) $\|\alpha v\| = |\alpha|\|v\|$, for all $\alpha \in \mathbb{R}$, $v \in V$,
(3) $\|v_1 + v_2\| \leq \|v_1\| + \|v_2\|$, for all $v_1, v_2 \in V$.

On $\mathbb{R}^n$, there are several norms. Suppose

$$x = \begin{bmatrix} x_1 \\ x_2 \\ \vdots \\ x_n \end{bmatrix} \in \mathbb{R}^n.$$

Some convenient norms are given by $\|x\| = \max_{1 \leq i \leq n} |x_i|$, or

$$\|x\| = \sum_{i=1}^n |x_i|, \quad \text{or} \quad \|x\| = \left(\sum_{i=1}^n x_i^2\right)^{\frac{1}{2}}.$$

On several occasions, we will also be dealing with linear transformations $A : \mathbb{R}^n \to \mathbb{R}^n$, which can be expressed by an $n \times n$ matrix whose action is determined by a set of basis elements. We will use the standard basis

$$e_1 = \begin{bmatrix} 1 \\ 0 \\ 0 \\ \vdots \\ 0 \\ 0 \end{bmatrix}, \quad e_2 = \begin{bmatrix} 0 \\ 1 \\ 0 \\ \vdots \\ 0 \\ 0 \end{bmatrix}, \quad \ldots, \quad e_n = \begin{bmatrix} 0 \\ 0 \\ 0 \\ \vdots \\ 0 \\ 1 \end{bmatrix},$$

and denote

$$A = \begin{bmatrix} a_{11} & a_{12} & \cdots & a_{1n} \\ a_{21} & a_{22} & \cdots & a_{2n} \\ \vdots & \vdots & \ddots & \vdots \\ a_{n1} & a_{n2} & \cdots & a_{nn} \end{bmatrix},$$

with matrix multiplication being the action of A on an n-vector.

Definition 1.3. Associated with such a transformation A is *a norm* defined by

$$\|A\| \equiv \inf \left\{ M > 0 \mid \|Ax\| \le M\|x\| \text{ for all } x \in \mathbb{R}^n \right\},$$
$$\text{or} \quad \|A\| \equiv \sup \left\{ \|Ax\| \mid \|x\| \le 1 \text{ for all } x \in \mathbb{R}^n \right\}.$$

These are equivalent definitions.

⬛ **Exercise** **2.** Take $\|x\| = \max_{1 \le i \le n} |x_i|$. Show that if $A : \mathbb{R}^n \to \mathbb{R}^n$ is as above, then its induced norm is given by

$$\|A\| = \max_{1 \le i \le n} \left(\sum_{j=1}^n |a_{ij}| \right).$$

1.3   The Picard Existence Theorem

We introduce some background material for this classical result.

Definition 1.4. The function $f(t, x)$ is said to satisfy *a Lipschitz condition* with respect to (wrt for short) x on $D \subseteq \mathbb{R} \times \mathbb{R}^n$ in case there is a constant $K > 0$ such that

$$\|f(t, x) - f(t, y)\| \le K\|x - y\|, \quad \text{for all } (t, x), (t, y) \in D.$$

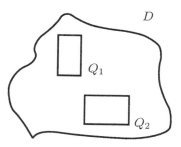

Fig. 1.1   Rectangles Q_1 and Q_2.

Definition 1.5. $f(t, x)$ is said to satisfy *a local Lipschitz condition* wrt x on $D \subseteq \mathbb{R} \times \mathbb{R}^n$ in case, corresponding to every "rectangle" $Q = \{(t, x) \mid |t - t_0| \le a, \|x - x_0\| \le b\} \subseteq D$, there is a positive $K_Q > 0$ such that

$$\|f(t, x) - f(t, y)\| \le K_Q \|x - y\|, \quad \text{for all } (t, x), (t, y) \in Q.$$

Note: As one goes from one rectangle to another, the constant K_Q varies, i.e., K_{Q_1} is not necessarily equal to K_{Q_2}. See Figure 1.1.

Lemma 1.1. *Let* $f(t, x)$ *and* $\frac{\partial f_i}{\partial x_j}$, $1 \le i, j \le n$, *be continuous on an open set* $D \subseteq \mathbb{R} \times \mathbb{R}^n$. *Then* f *satisfies a local Lipschitz condition wrt* x *on* D.

Proof. Let $f_x(t, x)$ denote the Jacobian matrix of $f(t, x)$; i.e.,

$$f_x(t, x) = \begin{bmatrix} \frac{\partial f_1}{\partial x_1} & \frac{\partial f_1}{\partial x_2} & \cdots & \frac{\partial f_1}{\partial x_n} \\ \frac{\partial f_2}{\partial x_1} & \frac{\partial f_2}{\partial x_2} & \cdots & \frac{\partial f_2}{\partial x_n} \\ \vdots & \vdots & \ddots & \vdots \\ \frac{\partial f_n}{\partial x_1} & \frac{\partial f_n}{\partial x_2} & \cdots & \frac{\partial f_n}{\partial x_n} \end{bmatrix}.$$

Now let $Q = \{(t, x) \mid |t - t_0| \le a, \|x - x_0\| \le b\} \subseteq D$ be a rectangle in D, and let $K_Q = \max_{(t,x) \in Q} \|f_x(t, x)\|$. Note here that, by the Exercise 2,

$$K_Q = \max_{(t,x) \in Q} \left[\max_{1 \le i \le n} \left(\sum_{j=1}^n \left| \frac{\partial f_i}{\partial x_j} \right| \right) \right].$$

We claim that f satisfies a Lipschitz condition wrt x on Q with Lipschitz coefficient K_Q. (By continuity, it is true that K_Q exists.)

For the claim, let $(t, x), (t, y) \in Q$ and set $z(s) = (1-s)x + sy$, $0 \le s \le 1$. We will first show that $(t, z(s)) \in Q$, for all $0 \le s \le 1$. Consider

$$
\begin{aligned}
\|z(s) - x_0\| &= \|(1-s)x + sy - x_0\| \\
&= \|(1-s)x + sy - (1-s)x_0 - sx_0\| \\
&= \|(1-s)(x - x_0) + s(y - x_0)\| \\
&\le (1-s)\|x - x_0\| + s\|y - x_0\| \\
&\le (1-s)b + sb = b.
\end{aligned}
$$

Since $|t - t_0| \le a$ by choice of t, we have $(t, z(s)) \in Q$, $0 \le s \le 1$.

(One might note here that we have shown Q to be a convex set wrt x.)

Above where we have $z(s) = (1-s)x + sy$, we mean of course that $z_1(s) = (1-s)x_1 + sy_1$, $z_2(s) = (1-s)x_2 + sy_2$, ..., $z_n(s) = (1-s)x_n + sy_n$, since $x, y \in \mathbb{R}^n$.

Thus, recalling the chain rule, $\frac{df_1}{ds}(t, z_1(s), \ldots, z_n(s)) = \frac{\partial f_1}{\partial x_1}\frac{dz_1}{ds} + \frac{\partial f_1}{\partial x_2}\frac{dz_2}{ds} + \cdots + \frac{\partial f_1}{\partial x_n}\frac{dz_n}{ds}$, etc., so that

$$
\frac{d}{ds}f(t, z(s)) = \begin{bmatrix} \frac{\partial f_1}{\partial x_1}(y_1 - x_1) + \frac{\partial f_1}{\partial x_2}(y_2 - x_2) + \cdots + \frac{\partial f_1}{\partial x_n}(y_n - x_n) \\ \vdots \\ \frac{\partial f_n}{\partial x_1}(y_1 - x_1) + \frac{\partial f_n}{\partial x_2}(y_2 - x_2) + \cdots + \frac{\partial f_n}{\partial x_n}(y_n - x_n) \end{bmatrix}_{(t, z(s))}
$$

$$
= f_x(t, z(s))(y - x).
$$

Hence,

$$
\int_0^1 \frac{d}{ds}f(t, z(s))\, ds = \int_0^1 f_x(t, z(s))(y - x)\, ds,
$$

which implies

$$
f(t, z(1)) - f(t, z(0)) = \int_0^1 f_x(t, z(s))(y - x)\, ds.
$$

But $z(1) = y$ and $z(0) = x$, so $f(t, y) - f(t, x) = \int_0^1 f_x(t, z(s))(y - x)\, ds$. [This is the Mean Value Theorem for vector-valued functions.]

Finally, we have

$$
\begin{aligned}
\|f(t, y) - f(t, x)\| &= \left\| \int_0^1 f_x(t, z(s))(y - x)\, ds \right\| \\
&\le \int_0^1 \|f_x(t, z(s))(y - x)\|\, ds \quad \text{(Can you verify this?)} \\
&\le \int_0^1 \|f_x(t, z(s))\| \cdot \|(y - x)\|\, ds \\
&\le \int_0^1 K_Q\|y - x\|\, ds = K_Q\|y - x\|.
\end{aligned}
$$

Therefore, f satisfies a local Lipschitz condition wrt x. □

Theorem 1.1 (Picard Existence Theorem). *Let $f(t,x)$ be continuous and satisfy a local Lipschitz condition wrt x on an open set $D \subseteq \mathbb{R} \times \mathbb{R}^n$. Let $(t_0, x_0) \in D$ and assume that the numbers $a, b > 0$ are such that the rectangle $Q = \{(t,x) \mid |t - t_0| \leq a, \|x - x_0\| \leq b\} \subseteq D$. Let $M = \max_{(t,x) \in Q} \|f(t,x)\|$ and let $\alpha = \min\left\{a, \frac{b}{M}\right\}$.*
Then the IVP,

$$\begin{cases} x' = f(t,x), \\ x(t_0) = x_0, \end{cases} \tag{1.1}$$

has a unique solution on $[t_0 - \alpha, t_0 + \alpha]$.

Proof. It suffices to show that there is a unique continuous n-vector function $\varphi(t)$ on $[t_0 - \alpha, t_0 + \alpha]$ such that $(t, \varphi(t)) \in Q \subseteq D$, for $|t - t_0| \leq \alpha$, and $\varphi(t) = x_0 + \int_{t_0}^{t} f(s, \varphi(s)) \, ds$, for $|t - t_0| \leq \alpha$.
Before proceeding with the proof, let's consider an example.

Example 1.1. Let $x' = t^2 + x^2 = f(t,x)$. Then, f is continuous on $D =$ the (t,x)-plane and f satisfies a local Lipschitz condition on the plane. Let us try to calculate the value of α.

Consider the IVP, $x' = t^2 + x^2$, $x(0) = 0$, and let $a, b > 0$ be fixed. Let the rectangle $Q = \{(t,x) \mid |t| \leq a, |x| \leq b\} \subseteq D =$ the (t,x)-plane. See Figure 1.2.

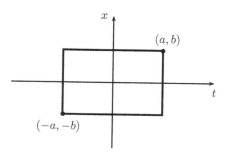

Fig. 1.2   Rectangle Q.

Now the $\max_{(t,x) \in Q} f(t,x)$ occurs at the corners of Q and thus $M = a^2 + b^2$, so that $\alpha = \min\left\{a, \frac{b}{a^2 + b^2}\right\}$.

For this example, we can determine the best possible (maximum) α. For fixed a, we find that the maximum of $\frac{b}{a^2 + b^2}$ as a function of b happens when $b = a$, so that $\max_b \frac{b}{a^2 + b^2} = \frac{a}{2a^2} = \frac{1}{2a}$. Thus $\alpha = \min\left\{a, \frac{1}{2a}\right\}$.

To find the maximum α, consider the decreasing function $s = \frac{1}{2a}$ and the increasing function $s = a$. The maximum α occurs when $a = \frac{1}{2a}$, or $a = \frac{1}{\sqrt{2}}$; i.e., we have $\alpha = \frac{1}{\sqrt{2}}$. See Figure 1.3.

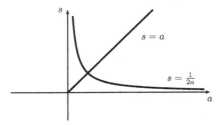

Fig. 1.3   Graphs of $s = a$ and $s = \frac{1}{2a}$.

Therefore, by the Picard Theorem, the IVP has a unique solution on $\left[-\frac{1}{\sqrt{2}}, \frac{1}{\sqrt{2}}\right]$.

We now continue the proof by showing that $\varphi(t) = x_0 + \int_{t_0}^{t} f(s, \varphi(s))\,ds$ has a unique continuous solution on $[t_0 - \alpha, t_0 + \alpha]$. Define a sequence $\{x_n(t)\}_{n=0}^{\infty}$, called the *sequence of Picard iterates*, by

$$x_0(t) = x_0,$$

$$x_n(t) = x_0 + \int_{t_0}^{t} f(s, x_{n-1}(s))\,ds, \ n \geq 1, \text{ on } [t_0 - \alpha, t_0 + \alpha].$$

We shall first prove that this sequence is well-defined by showing that $(t, x_n(t)) \in Q$, for $|t - t_0| \leq \alpha$ and $n \geq 0$. Clearly $(t, x_0(t)) = (t, x_0) \in Q$, for $|t - t_0| \leq \alpha \leq a$. We proceed by induction.

Assume that $(t, x_n(t)) \in Q$, for $|t - t_0| \leq \alpha$ and $0 \leq n \leq k-1$. Consider

$$x_k(t) = x_0 + \int_{t_0}^{t} f(s, x_{k-1}(s))\,ds.$$

Then

$$\|x_k(t) - x_0\| = \left\|\int_{t_0}^{t} f(s,\ x_{k-1}(s))\,ds\right\| \leq \left|\int_{t_0}^{t} \|f(s,\ x_{k-1}(s))\|\,ds\right|.$$

By the induction hypothesis, $(s, x_{k-1}(s)) \in Q$ implies $\|f(s, x_{k-1}(s))\| \leq M$, thus

$$\|x_k(t) - x_0\| \leq \left|\int_{t_0}^{t} M\,ds\right| \leq M|t - t_0| \leq M\alpha \leq b.$$

So, $(t, x_k(t)) \in Q$, for $|t - t_0| \leq \alpha$. Therefore, by induction, we have $(t, x_n(t)) \in Q$, for $|t - t_0| \leq \alpha$ and $n \geq 0$.

Now, let $K \geq 0$ be the Lipschitz coefficient for f on Q. Then we have $\|x_1(t) - x_0(t)\| = \left\|\int_{t_0}^t f(s, x_0(s))\, ds\right\| \leq M|t - t_0|$ on $[t_0 - \alpha,\, t_0 + \alpha]$, and

$$\|x_2(t) - x_1(t)\| = \left\|\int_{t_0}^t f(s, x_1(s)) - f(s, x_0(s))\, ds\right\|$$

$$\leq \left|\int_{t_0}^t \|f(s, x_1(s)) - f(s, x_0(s))\|\, ds\right|$$

$$\leq \left|\int_{t_0}^t K\|x_1(s) - x_0(s)\|\, ds\right|$$

$$\leq \left|\int_{t_0}^t KM|s - t_0|\, ds\right|$$

$$= \frac{KM|t - t_0|^2}{2}, \quad \text{on } [t_0 - \alpha, t_0 + \alpha].$$

We claim that $\|x_{n+1}(t) - x_n(t)\| \leq \frac{K^n M|t-t_0|^{n+1}}{(n+1)!}$ on $[t_0 - \alpha,\, t_0 + \alpha]$ for $n \geq 0$.

Again, we argue inductively. From above, it follows that the assertion is true for $n = 0,\, 1$. Assume the assertion is true for $0 \leq n = k - 1$ and consider

$$\|x_{k+1}(t) - x_k(t)\|$$

$$= \left\|\int_{t_0}^t (f(s,\, x_k(s)) - f(s,\, x_{k-1}(s)))\, ds\right\|$$

$$\leq \left|\int_{t_0}^t \|f(s,\, x_k(s)) - f(s,\, x_{k-1}(s))\|\, ds\right|$$

$$\leq \left|\int_{t_0}^t K\|x_k(s) - x_{k-1}(s)\|\, ds\right|$$

$$\leq \left|\int_{t_0}^t \frac{K\, K^{k-1}M|s - t_0|^k}{k!}\, ds\right| \quad \text{(by induction hypothesis)}$$

$$= \frac{K^k M}{k!}\left|\int_{t_0}^t |s - t_0|^k\, ds\right|$$

$$= \frac{K^k M|t - t_0|^{k+1}}{k!(k+1)}$$

$$= \frac{K^k M|t - t_0|^{k+1}}{(k+1)!}.$$

Therefore, by induction, $\|x_{n+1}(t) - x_n(t)\| \leq \frac{K^n M |t-t_0|^{n+1}}{(n+1)!}$ for $|t - t_0| \leq \alpha$ and $n \geq 0$. Thus, $\|x_{n+1}(t) - x_n(t)\| \leq \frac{K^n M \alpha^{n+1}}{(n+1)!}$ for $|t - t_0| \leq \alpha$ and $n \geq 0$.

Let us consider the infinite series given by the telescoping sum

$$x_0 + \sum_{j=0}^{\infty} [x_{j+1}(t) - x_j(t)], \tag{1.3}$$

whose nst partial sum is $S_{n-1} = x_0 + x_1 - x_0 + x_2 - x_1 + \cdots + x_{n-1} - x_{n-2} + x_n - x_{n-1} = x_n(t)$.

Now, $\|x_{j+1}(t) - x_j(t)\| \leq \frac{K^j M \alpha^{j+1}}{(j+1)!}$ and we also know that $\sum_{j=0}^{\infty} \frac{K^j M \alpha^{j+1}}{(j+1)!}$ converges (to what?).

Therefore, by the Weierstrass M-test, (1.3) converges uniformly on $[t_0 - \alpha, t_0 + \alpha]$; i.e., the sequence of partial sums $\{S_n\} = \{x_n(t)\}$ converges uniformly on $[t_0 - \alpha, t_0 + \alpha]$. So, let $\lim_{x \to \infty} x_n(t) := \varphi(t)$. Then, $\varphi(t)$ is continuous.

Now, since Q is a closed and bounded subset of $\mathbb{R} \times \mathbb{R}^n$, Q is compact. Hence f is uniformly continuous on Q, and as a consequence, $\{f(t, x_n(t))\}$ converges uniformly on $[t_0 - \alpha, t_0 + \alpha]$. Hence, from the fact that $x_n(t) = x_0 + \int_{t_0}^t f(s, x_{n-1}(s)) \, ds$, we can take limits as follows

$$\lim_{n \to \infty} x_n(t) = x_0 + \lim_{n \to \infty} \int_{t_0}^t f(s, x_{n-1}(s)) \, ds$$

$$= x_0 + \int_{t_0}^t \lim_{n \to \infty} f(s, x_{n-1}(s)) \, ds,$$

which is because of the uniform convergence of $\{f(t, x_n(t))\}$. Hence,

$$\varphi(t) = x_0 + \int_{t_0}^t f(s, \varphi(s)) \, ds.$$

Therefore, φ satisfies the appropriate integral equation and is consequently a solution of the desired IVP on $[t_0 - \alpha, t_0 + \alpha]$.

Before establishing the uniqueness of φ, we consider a number of asides.

Corollary 1.1. *In the iterative sequence* $\{x_n(t)\}$, *the error is bounded by* $\frac{K^n M \alpha^{n+1}}{(n+1)!} e^{K\alpha}$, $n = 0, 1, 2, \ldots$.

Proof. Since $\varphi(t) = \lim_{n \to \infty} x_n(t)$, we have

$$\varphi(t) = x_0 + \sum_{j=0}^{\infty} [x_{j+1}(t) - x_j(t)].$$

Now

$$x_n(t) = x_0 + \sum_{j=0}^{n-1} [x_{j+1}(t) - x_j(t)],$$

and so

$$\|\varphi(t) - x_n(t)\| \le \sum_{j=n}^{\infty} \|x_{j+1}(t) - x_j(t)\| \le \sum_{j=n}^{\infty} \frac{K^j M \alpha^{j+1}}{(j+1)!}$$

$$= \frac{K^n M \alpha^{n+1}}{(n+1)!} \left(1 + \sum_{j=n+1}^{\infty} \frac{K^{j-n} \alpha^{j-n}}{(n+2)\cdots(j+1)} \right)$$

$$\le \frac{K^n M \alpha^{n+1}}{(n+1)!} e^{K\alpha}. \qquad \square$$

Example 1.2. Let us compute this error for the IVP: $x' = t^2 + x^2$, $x(0) = 0$.

Take $D = \mathbb{R} \times \mathbb{R}$. Previously, we found that a unique solution exists on $\left[-\frac{1}{\sqrt{2}}, \frac{1}{\sqrt{2}}\right]$. In this case our local Lipschitz coefficient is determined by $\frac{\partial f}{\partial x} = 2x$, which yields

$$K_Q = \max_Q |2x| = 2\left(\frac{1}{\sqrt{2}}\right) = \sqrt{2},$$

where we are taking

$$Q = \left[-\frac{1}{\sqrt{2}}, \frac{1}{\sqrt{2}}\right] \times \left[-\frac{1}{\sqrt{2}}, \frac{1}{\sqrt{2}}\right]$$

from the maximizing process for α. Also, $M = \max_Q \|f(t,x)\| = \max_Q |t^2 + x^2| = \frac{1}{2} + \frac{1}{2} = 1$. Thus, for any $n = 0, 1, 2, \ldots,$

$$\|\varphi(t) - x_n(t)\| \le \frac{\sqrt{2}^n \cdot 1 \cdot \left(\frac{1}{\sqrt{2}}\right)^{n+1} \cdot e^{\sqrt{2} \cdot \frac{1}{\sqrt{2}}}}{(n+1)!} = \frac{e}{\sqrt{2}(n+1)!}.$$

Exercise 3. For each of the following, determine the best possible α, the corresponding Lipschitz coefficient K, and calculate the first 3 Picard iterates.

(1) $y' = y^3$, $y(0) = 2$.

(2) $y' = x + xy^2$, $y(0) = 3$ ($\alpha = 0.28$).

(3) $\begin{cases} y_1' = y_1 + y_2, & y_1(1) = -1, \\ y_2' = y_1 - y_2, & y_2(1) = 1. \end{cases}$

Before completing the proof of Theorem 1.1, we consider the following theorem first:

Theorem 1.2 (Gronwall Inequality). *Let* f, g *be continuous nonnegative real-valued functions defined on an interval* $I \subseteq \mathbb{R}$. *Let* $K(t)$ *be a continuous nonnegative function on* I, *and assume further that, there is a point* $t_0 \in I$ *such that*

$$f(t) \le g(t) + \left| \int_{t_0}^t K(s)f(s)\,ds \right|, \quad for\ all\ t \in I. \tag{1.4}$$

Then,

$$f(t) \le g(t) + \left| \int_{t_0}^t K(s)g(s)e^{\left| \int_s^t K(r)\,dr \right|}\,ds \right|, \quad for\ all\ t \in I. \tag{1.5}$$

Proof. Case I: Let $t \in I$ with $t > t_0$ be fixed. Then by (1.4), $f(u) \le g(u) + \int_{t_0}^u K(s)f(s)\,ds$, for all $t_0 \le u \le t$. If we set $\psi(u) \equiv \int_{t_0}^u K(s)f(s)\,ds$, then ψ is differentiable and $\psi'(u) = K(u)f(u)$.

Since $K(u) \ge 0$, we have

$$K(u)f(u) \le K(u)g(u) + K(u)\int_{t_0}^u K(s)f(s)\,ds,$$

which is

$$K(u)f(u) \le K(u)g(u) + K(u)\psi(u) \quad \text{for } t_0 \le u \le t,$$

or

$$\psi'(u) \le K(u)g(u) + K(u)\psi(u) \quad \text{for } t_0 \le u \le t.$$

Rewriting as $\psi'(u) - K(u)\psi(u) \le K(u)g(u)$, and then multiplying both sides of this latter inequality by the integrating factor $e^{-\int_{t_0}^u K(r)\,dr} > 0$, we have

$$e^{-\int_{t_0}^u K(r)\,dr}[\psi'(u) - K(u)\psi(u)] \le e^{-\int_{t_0}^u K(r)\,dr}K(u)g(u),$$

or

$$[e^{-\int_{t_0}^u K(r)\,dr}\psi(u)]' \le e^{-\int_{t_0}^u K(r)\,dr}K(u)g(u).$$

Now integrate over $[t_0, t]$ and recall both sides are nonnegative, and we have

$$e^{-\int_{t_0}^u K(r)\,dr}\psi(u)|_{t_0}^t \le \int_{t_0}^t e^{-\int_{t_0}^u K(r)\,dr}K(u)g(u)\,du,$$

which gives

$$e^{-\int_{t_0}^t K(r)\,dr}\psi(t) \le \int_{t_0}^t e^{-\int_{t_0}^u K(r)\,dr} K(u)g(u)\,du,$$

so that

$$\psi(t) \le e^{\int_{t_0}^t K(r)\,dr} \int_{t_0}^t e^{-\int_{t_0}^u K(r)\,dr} K(u)g(u)\,du$$

$$= \int_{t_0}^t e^{\int_u^t K(r)\,dr} K(u)g(u)\,du.$$

Therefore, $f(t) \le g(t) + \psi(t) \le g(t) + \int_{t_0}^t e^{\int_u^t K(r)\,dr} K(u)g(u)\,du$, for all $t \in I$ such that $t > t_0$.

$\boxed{\textbf{Exercise}}$ **4.** Verify equation (1.5) when $t < t_0$.

This completes the proof of this theorem. $\qquad\square$

Corollary 1.2. *If the hypotheses of Theorem 1.2 are satisfied, but $g(t) \equiv 0$ on I, then $f \equiv 0$ on I.*

Proof. By (1.5) in Theorem 1.2, we have

$$f(t) \le g(t) + \left| \int_{t_0}^t K(s)g(s)\, e^{\left|\int_s^t k(r)\,dr\right|}\,ds \right| = 0 + 0 = 0.$$

Since f is nonnegative on I, we have $f \equiv 0$ on I. $\qquad\square$

Proof of Theorem 1.1 continued: We now establish the uniqueness of the solution φ to the IVP (1.1) on $[t_0 - \alpha, t_0 + \alpha]$. What we mean by this is as follows: Let $\varphi(t)$ be the solution we obtained above on $[t_0 - \alpha, t_0 + \alpha]$, and let $\psi(t)$ be a solution to the same IVP, $x' = f(t, x)$, $x(t_0) = x_0$, on an interval J containing t_0. Then we will show that $\varphi(t) \equiv \psi(t)$ on $[t_0 - \alpha, t_0 + \alpha] \cap J$.

For such a solution $\psi(t)$, we **claim** that $\|\psi(t) - x_0\| \le b$ for all $t \in [t_0 - \alpha, t_0 + \alpha] \cap J$.

Assume this is not true, i.e., $(t, \psi(t)) \notin Q$ for some $t \in [t_0 - \alpha, t_0 + \alpha] \cap J$. Then $\|\psi(t) - x_0\| > b$ for some $t \in [t_0 - \alpha, t_0 + \alpha] \cap J$. However, $x_0 = \psi(t_0)$, and hence by continuity, there exists $\tau \in (t_0 - \alpha, t_0 + \alpha) \cap J$ such that $\|\psi(\tau) - x_0\| = b$ and $\|\psi(t) - x_0\| < b$ on $[t_0, \tau)$ or on $(\tau, t_0]$.

Now $\psi(\tau) = x_0 + \int_{t_0}^\tau f(s, \psi(s))\,ds$ and so

$$b = \|\psi(\tau) - x_0\| \le \left| \int_{t_0}^\tau \|f(s, \psi(s))\|\,ds \right| \le M|\tau - t_0| < M\alpha \le b,$$

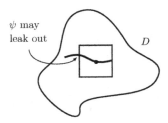

Fig. 1.4   The solution $\psi(t)$.

which is a contradiction. Therefore, $(t, \psi(t)) \in Q$ for all $t \in [t_0 - \alpha, t_0 + \alpha] \cap J$. Hence for any such t, we have $\varphi(t) = x_0 + \int_{t_0}^{t} f(s, \varphi(s)) \, ds$ and $\psi(t) = x_0 + \int_{t_0}^{t} f(s, \psi(s)) \, ds$, and so,

$$\varphi(t) - \psi(t) = \int_{t_0}^{t} [f(s, \varphi(s)) - f(s, \psi(s))] \, ds,$$

which yields

$$\|\varphi(t) - \psi(t)\| \leq \left| \int_{t_0}^{t} \|f(s, \varphi(s)) - f(s, \psi(s))\| \, ds \right|$$

$$\leq 0 + \left| \int_{t_0}^{t} K \|\varphi(s) - \psi(s)\| \, ds \right|.$$

By Corollary 1.2 to Theorem 1.2, it follows that $\|\varphi(t) - \psi(t)\| = 0$, that is, $\varphi(t) = \psi(t)$ for all $t \in [t_0 - \alpha, t_0 + \alpha] \cap J$.

Therefore, φ is the unique solution and the proof of Theorem 1.1 is complete. $\square$

Theorem 1.3. *Let $f(t, x)$ be continuous on $D = I \times \mathbb{R}^n$, where I is an interval of the reals. Assume that f satisfies the Lipschitz condition wrt x on each subset in $\mathbb{R} \times \mathbb{R}^n$ of the form $[a, b] \times \mathbb{R}^n$, where $[a, b]$ (compact) $\subseteq I$. Then, for any $t_0 \in I$ and any $x_0 \in \mathbb{R}^n$, the IVP $x' = f(t, x)$, $x(t_0) = x_0$ has a unique solution on I.*

Proof. Let a fixed IVP as specified be given and let $\tau \in I$ be fixed, but arbitrary. Choose a compact interval $[a, b] \subseteq I$ such that $t_0, \tau \in [a, b]$.

Define a sequence $\{x_n(t)\}$ by

$$x_0(t) = x_0, \quad x_n(t) = x_0 + \int_{t_0}^{t} f(s, x_{n-1}(s)) \, ds.$$

Then,

$$x_1(t) - x_0(t) = \int_{t_0}^t f(s, x_0(s))\, ds = \int_{t_0}^t f(s, x_0)\, ds.$$

If $M = \max_{a \le t \le b} \|f(t, x_0)\|$, then $\|x_1(t) - x_0(t)\| \le M|t - t_0|$ on $[a, b]$.

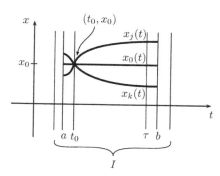

Fig. 1.5 The sequence $x_n(t)$.

As in Theorem 1.1, we have $\|x_{n+1}(t) - x_n(t)\| \le \frac{MK^n|t-t_0|^{n+1}}{(n+1)!}$, for all $n \ge 0$, $t \in [a, b]$, where K is the Lipschitz coefficient of f on $[a, b] \times \mathbb{R}^n$. In particular, $\|x_{n+1}(t) - x_n(t)\| \le \frac{M K^n|b-a|^{n+1}}{(n+1)!}$. As in Theorem 1.1, the sequence $\{x_n(t)\}$ will converge uniformly to a solution $\varphi(t)$ of the IVP on $[a, b]$. Since τ was chosen arbitrarily, φ is defined on the entire interval I. Moreover, φ is unique from the Gronwall inequality, i.e., given another solution $\psi(t)$ of the IVP and an arbitrary compact $[a, b] \subseteq I$, with $t_0 \in [a, b]$, $\|\varphi(t) - \psi(t)\| \le |\int_{t_0}^t K\|\varphi(s) - \psi(s)\|\, ds|$ on $[a, b]$. By Corollary 1.2 to Theorem 1.2, $\|\varphi(t) - \psi(t)\| \equiv 0$ on $[a, b]$. Hence, $\varphi(t) \equiv \psi(t)$ on $[a, b]$. Since $[a, b]$ was arbitrary, $\varphi(t) \equiv \psi(t)$ on I. $\qquad\square$

Example 1.3 (Where Theorem 1.3 is applicable). Consider the linear system

$$\begin{aligned}
x_1' &= a_{11}(t)x_1 + a_{12}(t)x_2 + \cdots + a_{1n}(t)x_n + h_1(t), \\
x_2' &= a_{21}(t)x_1 + a_{22}(t)x_2 + \cdots + a_{2n}(t)x_n + h_2(t), \\
&\;\;\vdots \\
x_n' &= a_{n1}(t)x_1 + a_{n2}(t)x_2 + \cdots + a_{nn}(t)x_n + h_n(t),
\end{aligned}$$

where a_{ij}, h_i, $1 \le i, j \le n$, are real-valued continuous functions defined on an interval $I \subseteq \mathbb{R}$.

Rewrite the system as

$$x' = A(t)x + h(t) := f(t, x).$$

Clearly f is continuous on $I \times \mathbb{R}^n$. If $[a, b]$ (compact) $\subseteq I$, consider $\|f(t, x) - f(t, y)\| = \|A(t)(x - y)\| \leq K\|x - y\|$, for $(t, x), (t, y) \in [a, b] \times \mathbb{R}^n$, where $K = \max_{t \in [a,b]} \|A(t)\|$. If we take vector norm $\|x\| = \max_{1 \leq j \leq n} |x_j|$, then

$$K = \max_{t \in [a,b]} \left\{ \max_{1 \leq i \leq n} \left\{ \sum_{1 \leq i \leq n} |a_{ij}(t)| \right\} \right\}.$$

The hypotheses of Theorem 1.3 are satisfied, thus the theorem can be applied.

Before we consider a second example, we will look at an nth order scalar equation and its translation to a first order system.

Consider

$$x^{(n)} = f(t, x, x', \ldots, x^{(n-1)}), \tag{1.6}$$

where f and x are real-valued functions. Define

$$
\begin{aligned}
y_1 &= x, \\
y_2 &= x' = y_1', \\
y_3 &= x'' = y_2', \\
&\vdots \\
y_{n-1} &= x^{(n-2)} = y_{n-2}', \\
y_n &= x^{(n-1)} = y_{n-1}', \\
(&\Rightarrow \quad x^{(n)} = y_n'),
\end{aligned}
\implies
\begin{cases}
y_1' = y_2, \\
y_2' = y_3, \\
\quad \vdots \\
y_{n-1}' = y_n, \\
y_n' = f(t, y_1, y_2, \ldots, y_n),
\end{cases}
\tag{1.7}
$$

or

$$y' = \tilde{f}(t, y), \text{ where } y = \begin{bmatrix} y_1 \\ y_2 \\ \vdots \\ y_n \end{bmatrix}, \quad \tilde{f}(t, y) = \begin{bmatrix} y_2 \\ y_3 \\ \vdots \\ y_n \\ f(t, y_1, y_2, \ldots, y_n) \end{bmatrix}. \tag{1.8}$$

Thus, we see that if $x(t) \in C^{(n)}(I)$ is a solution of (1.6) on I, then

$$y(t) = \begin{bmatrix} y_1(t) \\ \vdots \\ y_n(t) \end{bmatrix} = \begin{bmatrix} x(t) \\ \vdots \\ x^{(n-1)}(t) \end{bmatrix}$$

is a solution of (1.7) or (1.8) on I with $y \in C^{(1)}(I)$.

Conversely, if $y(t) \in C^{(1)}(I)$ is a solution of (1.7) or (1.8) on I, then $y_1(t)$ is a solution of (1.6) on I.

We can write an IVP for (1.8) as follows:

$$\begin{cases} y' = \tilde{f}(t, y), \\ y(t_0) = C, \quad \text{where } C = \begin{bmatrix} C_1 \\ \vdots \\ C_n \end{bmatrix}. \end{cases}$$

This is equivalent to the IVP for (1.6):

$$\begin{cases} x^{(n)} = f(t, x, x', \dots, x^{(n-1)}), \\ x^{(i)}(t_0) = C_{i+1}, \quad 0 \le i \le n-1. \end{cases}$$

Example 1.4. We now consider a second example where Theorem 1.3 is applicable. Suppose we have the linear equation $x^{(n)}(t) = \sum_{i=0}^{n-1} p_i(t) x^{(i)}(t) + h(t)$, where p_i, h are real-valued and continuous on an interval $I \subseteq \mathbb{R}$.

Translate to the first order system by letting $y_1 = x$:

$$\begin{cases} y_1' = y_2(= x'), \\ y_2' = y_3(= x''), \\ \quad \vdots \\ y_{n-1}' = y_n(= x^{(n-1)}), \\ y_n' = \sum_{i=0}^{n-1} p_i(t) y_{i+1} + h(t), \end{cases}$$

which we rewrite as $y' = A(t)y + \tilde{h}(t)$, where

$$A(t) = \begin{bmatrix} 0 & 1 & 0 & 0 & \cdots & 0 \\ 0 & 0 & 1 & 0 & \cdots & 0 \\ \vdots & \vdots & \vdots & \vdots & \ddots & \vdots \\ 0 & 0 & 0 & 0 & \cdots & 1 \\ p_0 & p_1 & p_2 & p_3 & \cdots & p_{n-1} \end{bmatrix}, \quad \tilde{h}(t) = \begin{bmatrix} 0 \\ 0 \\ \vdots \\ 0 \\ h(t) \end{bmatrix}.$$

Referring to Example 1.3, for any $t_0 \in I$ and any c_i, $1 \leq i \leq n$, the IVP

$$\begin{cases} y' = A(t)y + \tilde{h}(t), \\ y(t_0) = c = \begin{bmatrix} c_1 \\ c_2 \\ \vdots \\ c_n \end{bmatrix} \end{cases}$$

has a unique solution by Theorem 1.3. Equivalently, the IVP

$$\begin{cases} x^{(n)}(t) = \sum_{i=0}^{n-1} p_i(t)x^{(i)}(t) + h(t), \\ x^{(i)}(t_0) = c_{i+1}, \ 0 \leq i \leq n-1 \end{cases}$$

has a unique solution $x(t) \in C^{(n)}(I)$.

$\boxed{\text{Exercise}}$ **5.** For the following second order scalar equations, show that Theorem 1.3 applies to yield unique $C^{(2)}$ solutions on $\mathbb{R}$, and calculate their solutions.

(1) $\begin{cases} x'' + |x| = 0, \\ x(0) = 0, \ x'(0) = 1. \end{cases}$

(2) $\begin{cases} x'' - 2|x'| + x = 0, \\ x(0) = 1, \ x'(0) = 0. \end{cases}$

1.4   The Peano Existence Theorem

In the Picard existence theorem, the fact that $f(t, x)$ was locally Lipschitz was instrumented in establishing the uniqueness as well as the existence of the solution φ.

Theorem 1.4 (Peano Existence Theorem). *Let $f(t, x)$ be continuous on the open set $D \subseteq \mathbb{R} \times \mathbb{R}^n$, let $(t_0, x_0) \in D$, and let $a, b > 0$ be such that the rectangle $Q = \{(t, x) \mid \|t - t_0\| \leq a, \|x - x_0\| \leq b\} \subseteq D$. Let $M = \max_{(t,x)\in Q}\|f(t, x)\|$ and $\alpha = \min\left\{a, \frac{b}{M}\right\}$. Then the IVP*

$$\begin{cases} x' = f(t, x), \\ x(t_0) = x_0 \end{cases}$$

has a solution on $[t_0 - \alpha, t_0 + \alpha]$.

Proof. We will prove the existence of a solution on $[t_0, t_0 + \alpha]$. In a similar way, one can prove the existence of a solution on $[t_0 - \alpha, t_0]$. Then the solutions fit together at t_0 to give a solution on $[t_0 - \alpha, t_0 + \alpha]$.

We now construct a sequence of approximate solutions which consists of polygonal lines, as follows:

For any integer $n \geq 1$, partition $[t_0, t_0 + \alpha]$ into 2^n equal subintervals with partition points $t_j = t_0 + \frac{j\alpha}{2^n}$, $0 \leq j \leq 2^n$.

Define $x_n(t)$ by

$$x_n(t) = x_0 + f(t_0, x_0)(t - t_0), \ t_0 \leq t \leq t_1,$$
$$x_n(t) = x_n(t_1) + f(t_1, x_n(t_1))(t - t_1), \ t_1 \leq t \leq t_2,$$
$$\vdots$$
$$x_n(t) = x_n(t_j) + f(t_j, x_n(t_j))(t - t_j), \ t_j \leq t \leq t_{j+1}, \tag{1.9}$$
$$\vdots$$

where

$$x_n(t_1) = x_0 + f(t_0, x_0)(t_1 - t_0),$$
$$x_n(t_2) = x_n(t_1) + f(t_1, x_n(t_1))(t_2 - t_1),$$
$$\vdots$$
$$x_n(t_j) = x_n(t_{j-1}) + f(t_{j-1}, x_n(t_{j-1}))(t_j - t_{j-1}), \tag{1.10}$$
$$\vdots$$

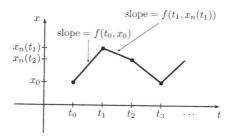

Fig. 1.6   The graph of $x_n(t)$.

The remainder of the proof consists of showing that a subsequence of $\{x_n(t)\}$ converges to a solution of the IVP on $[t_0, t_0 + \alpha]$.

Claim 1. *For all $n \geq 1$, $\{(t, x_n(t))\} \subseteq Q$, $t_0 \leq t \leq t_0 + \alpha$; i.e., for all $n \geq 1$, $\|x_n(t) - x_0\| \leq b$ on $[t_0, t_0 + \alpha]$.*

Proof of Claim 1. On the first subinterval, $t_0 \leq t \leq t_1$, we have that $\|x_n(t) - x_0\| = \|f(t_0, x_0)(t - t_0)\| \leq M(t_1 - t_0) \leq M\alpha \leq$ b. Thus the claim is satisfied on the first subinterval.

The proof proceeds by induction on j. Thus, assume now that $\|x_n(t) - x_0\| \leq b$ for $t_0 \leq t \leq t_j$ and consider $t \in [t_j, t_{j+1}]$.

Well,

$$\|x_n(t) - x_0\|$$
$$= \|x_n(t) - x_n(t_j) + x_n(t_j) - x_n(t_{j-1}) + x_n(t_{j-1}) - \cdots - x_0\|$$
$$\leq \|x_n(t) - x_n(t_j)\| + \sum_{k=1}^{j} \|x_n(t_k) - x_n(t_{k-1})\|$$
$$\leq \|f(t_j, x_n(t_j))\|(t - t_j)$$
$$+ \sum_{k=1}^{j} \|f(t_{k-1}, x_n(t_{k-1}))\|(t_k - t_{k-1}), \text{ (by (1.9), (1.10))}$$
$$\leq M(t - t_0) \leq M\alpha \leq b, \text{ (by induction hypothesis)}.$$

Therefore, Claim 1 is true for $t_0 \leq t \leq t_{j+1}$ and hence the claim follows by induction.   □

Claim 2. *For all $n \geq 1$ and any $t_0 \leq \tau \leq t \leq t_0 + \alpha$, $\|x_n(t) - x_n(\tau)\| \leq M|t - \tau|$; i.e., x_n satisfies a Lipschitz condition.*

Proof of Claim 2. (a) Assume $t_j \leq \tau \leq t \leq t_{j+1}$, for some j. From (1.9), we have

$$x_n(t) - x_n(\tau) = f(t_j, x_n(t_j))(t - \tau).$$

So,

$$\|x_n(t) - x_n(\tau)\| \leq \|f(t_j, x_n(t_j))\||t - \tau| \leq M|t - \tau|.$$

(b) Assume $t_j \leq \tau \leq t_{j+1} \leq \cdots \leq t_k \leq t \leq t_{k+1}$, for some j and k. Then

$$\|x_n(t) - x_n(\tau)\|$$
$$\leq \|x_n(t) - x_n(t_k)\| + \|x_n(t_{j+1}) - x_n(\tau)\|$$
$$+ \sum_{r=j+2}^{k} \|x_n(t_r) - x_n(t_{r-1})\|$$
$$\leq M|t - t_k| + M|t_{j+1} - \tau| + M \sum_{r=j+2}^{k} |t_r - t_{r-1}| \quad \text{(by (1.9) and (1.10))}$$
$$= M|t - \tau|.$$

Therefore, Claim 2 is true. □

From Claim 1, our sequence of polygonal lines is uniformly bounded; i.e., $\|x_n(t)\| \leq \|x_0\| + b$ on $[t_0, t_0 + \alpha]$. From Claim 2, the sequence is equicontinuous on $[t_0, t_0 + \alpha]$. Consequently, by the Ascoli-Arzelá Theorem, there exists a subsequence $\{x_{n_k}(t)\}$ which converges uniformly on $[t_0, t_0 + \alpha]$, say

$$\lim_{n_k \to \infty} x_{n_k}(t) = \varphi(t).$$

Note here that since $f(t, x)$ is continuous on Q (compact) $\subseteq \mathbb{R} \times \mathbb{R}^n$, we have that $f(t, x)$ is uniformly continuous on Q. Hence, for each $\varepsilon > 0$, there exists a $\delta > 0$ such that, for $(t, x), (\tau, y) \in Q$ with $|t - \tau| < \delta$, $\|x - y\| < \delta$ we have $\|f(t, x) - f(\tau, y)\| < \varepsilon$. It follows that, if n_k is large enough, say $\max\{\frac{\alpha}{2^{n_k}}, \frac{M\alpha}{2^{n_k}}\} < \delta$, then for t arbitrary, $t_j \leq t \leq t_{j+1}$, we have $|t - t_j| \leq |t_j - t_{j+1}| \leq \frac{\alpha}{2^{n_k}} < \delta$, and

$$\|x_{n_k}(t) - x_{n_k}(t_j)\| \leq \|f(t_j, x_{n_k}(t_j))\|(t - t_j) \leq \frac{M\alpha}{2^{n_k}} < \delta.$$

Hence, for all $t_j \leq t \leq t_{j+1}$ and $j = 0, 1, \ldots, 2^n$,

$$\|f(t, x_{n_k}(t)) - f(t_j, x_{n_k}(t_j))\| < \varepsilon. \tag{1.11}$$

But by (1.9),

$$x_{n_k}(t) = x_{n_k}(t_j) + f(t_j, x_{n_k}(t_j))(t - t_j), \text{ for } t_j \leq t \leq t_{j+1}.$$

Therefore,

$$x'_{n_k}(t) = 0 + f(t_j, x_{n_k}(t_j)), \text{ on } t_j \leq t \leq t_{j+1}. \tag{1.12}$$

From the construction of our polygonal line approximations, for $t_0 \leq t \leq t_0 + \alpha$, we have

$$x_{n_k}(t) = x_0 + \int_{t_0}^{t} x'_{n_k}(s) \, ds$$

$$= x_0 + \int_{t_0}^{t} \left[f(s, x_{n_k}(s)) + x'_{n_k}(s) - f(s, x_{n_k}(s)) \right] ds,$$

which yields

$$\left\| x_{n_k}(t) - x_0 - \int_{t_0}^{t} f(s, x_{n_k}(s)) \, ds \right\|$$

$$\leq \left| \int_{t_0}^{t} \left\| x'_{n_k}(s) - f(s, x_{n_k}(s)) \right\| ds \right|$$

$$\leq \varepsilon \alpha, \text{ for } n_k \text{ large, by (1.11) and (1.12).}$$

But $\lim_{n_k \to \infty} x_{n_k}(t) = \varphi(t)$ uniformly, which implies

$$\lim_{n_k \to \infty} \int_{t_0}^t f(s, x_{n_k}(s))\, ds = \int_{t_0}^t f(s, \varphi(s))\, ds.$$

So,

$$\lim_{n_k \to \infty} \left\| x_{n_k}(t) - x_0 - \int_{t_0}^t f(s, x_{n_k}(s))\, ds \right\|$$
$$= \left\| \varphi(t) - x_0 - \int_{t_0}^t f(s, \phi(s))\, ds \right\| = 0.$$

Therefore, $\varphi(t) = x_0 + \int_{t_0}^t f(s, \varphi(s))\, ds$, and $\varphi(t)$ is a solution to the IVP. $\square$

Example 1.5. This example illustrates that in the absence of a Lipschitz condition, solutions are not necessarily unique.

Consider the scalar equation $x' = \sqrt{|x|}$, $x(t_0) = 0$. Consider $t \geq t_0$ and the case $x(t) \geq 0$:

$$x' = \sqrt{x}, \quad t \geq t_0,$$
$$\implies \quad x^{-\frac{1}{2}} x' = 1, \quad t \geq t_0,$$
$$\implies \quad \int_{t_0}^t x^{-\frac{1}{2}}(t) x'(t)\, dt = \int_{t_0}^t dt, \quad t \geq t_0,$$
$$\implies \quad x(t) = \frac{1}{4}(t - t_0)^2, \quad t \geq t_0,$$

is a solution. Yet $x(t) \equiv 0$ is also a solution.

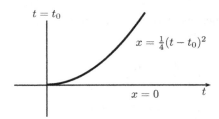

Fig. 1.7   Nonunique solutions.

Exercise **6.** Let $f(t, x)\colon \mathbb{R} \times \mathbb{R} \to \mathbb{R}$ be defined by $f(t, x) = \sqrt{|x|}$. Show that f does not satisfy a Lipschitz condition on any rectangle of the form $Q = \{(t, x) \mid |t - t_0| \leq a, |x - 0| \leq b\}$.

Remark 1.1. The IVP: $x' = \sqrt{|x|}$, $x(t_0) = 0$ has an infinite number of solutions.

Exercise 7. Verify Remark 1.1, by finding an infinite number of solutions of the IVP: $x' = \sqrt{|x|}$, $x(t_0) = 0$.

Corollary 1.3 (To Theorem 1.4). *Let $D \subseteq \mathbb{R} \times \mathbb{R}^n$ be open and let $f : D \to \mathbb{R}^n$ be continuous. Then if $K \subseteq D$ is compact, then there exists a $\delta_K > 0$ such that for all $(t_0, x_0) \in K$, the IVP $x' = f(t, x)$, $x(t_0) = x_0$ has a solution on $[t_0 - \delta_K, t_0 + \delta_K]$.*

Proof. Let $\eta = 1$, if $D = \mathbb{R} \times \mathbb{R}^n$, and let $\eta = \frac{1}{2}d(K, \partial D)$ if $D \neq \mathbb{R} \times \mathbb{R}^n$. Here d is given as follows: If $(t_1, x_1), (t_2, x_2) \in \mathbb{R} \times \mathbb{R}^n$, then $d((t_1, x_1), (t_2, x_2)) = |t_1 - t_2| + \|x_1 - x_2\|$.

Let $H = \{(t, x) | d((t, x), K) \leq \eta\}$. Then H(compact) $\subseteq D$. Let $M = \max_{(t,x) \in H} \|f\|$ and let $(t_0, x_0) \in K$. Then define $Q = \{(t, x) | |t - t_0| \leq \frac{1}{2}\eta, \|x - x_0\| \leq \frac{1}{2}\eta\}$.

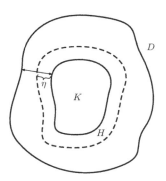

Fig. 1.8 $K \subseteq H \subseteq D$.

Note: If $(\hat{t}, \hat{x}) \in Q$, then $d((\hat{t}, \hat{x}), (t_0, x_0)) = |\hat{t} - t_0| + \|\hat{x} - x_0\| \leq \eta$.

Since $(t_0, x_0) \in K$, we have $d((\hat{t}, \hat{x}), K) \leq \eta$ and so $(\hat{t}, \hat{x}) \in H$. Thus, $Q \subseteq H \subset D$. (Note: Q is the type of "Q-rectangle" in Theorem 1.4, where $a = \frac{1}{2}\eta$, $b = \frac{1}{2}\eta$.)

Let $\delta_K = \min\left\{a, \frac{b}{M}\right\} = \left\{\frac{\eta}{2}, \frac{\eta}{2M}\right\}$. Then δ_K corresponds to the number α in Theorem 1.4; consequently, the hypotheses of Theorem 1.4 are satisfied and it follows that the IVP

$$\begin{cases} x' = f(t, x), \\ x(t_0) = x_0, \end{cases}$$

has a solution on $[t_0 - \delta_K, t_0 + \delta_K]$. $\square$

Before our final theorem in this section, we will look at some examples of Picard iterates of functions not satisfying Lipschitz conditions. In the first case, it will be true that the Picard iterates converge to desired solutions, whereas, in the second case the Picard iterates will not converge.

Example 1.6. Consider the IVP

$$\begin{cases} x' = \sqrt{|x|}(= f(t, x)), \\ x(0) = 0. \end{cases}$$

Obviously, $f(t, x) = \sqrt{|x|}$ does not satisfy a Lipschitz condition on any rectangle containing $x = 0$. The Picard iterates are given by

$$x_0(t) = x(0) = 0,$$

$$x_n(t) = \int_0^t |x_{n-1}(s)|^{\frac{1}{2}} ds = 0, \quad n \geq 1.$$

Hence, all Picard iterates are zero, which converge to a solution $\varphi \equiv 0$.

Note that in the proof of the Picard Theorem, one doesn't have to set $x_0(t) = x_0$. The only real restriction is that $\{(t, x_0(t))\} \subseteq Q$.

Hence, again consider the above IVP: $x' = \sqrt{|x|}$, $x(0) = 0$. We can define $x_0(t) = t^\alpha$, $\alpha > 0$ fixed, on $[0, \infty)$:

Exercise **8.** Show that $x_n(t) = C_n(\alpha) t^{e_n(\alpha)}$, where $e_n(\alpha) = \sum_{j=0}^{n-1} \frac{1}{2^j} + \frac{\alpha}{2^n}$ and $C_{n+1}(\alpha) = \frac{\sqrt{C_n(\alpha)}}{1 + \frac{1}{2} e_n(\alpha)}$, where $C_1(\alpha) = \frac{2}{2+\alpha}$.

Then $\lim_{n \to \infty} e_n(\alpha) = \lim_{n \to \infty} \sum_{j=0}^{n-1} \frac{1}{2^j} + \frac{\alpha}{2^n} = \sum_{j=0}^{\infty} \frac{1}{2^j} + 0 = 2$, and if $b = \lim_{n \to \infty} C_n(\alpha)$, then $b = \frac{\sqrt{b}}{1 + \frac{1}{2}(2)}$ implying $b = \lim_{n \to \infty} C_n(\alpha) = \frac{1}{4}$.

Hence, $\lim_{n \to \infty} x_n(t) = \frac{1}{4} t^2$ which is also a solution of the IVP.

Thus, we have an example of an IVP with no Lipschitz condition, yet the Picard iterates converge to two of the solutions, (recall, no unique solution).

Example 1.7. For this example, we consider a problem which does have a unique solution, however there is no Lipschitz condition — moreover, the Picard iterates do not converge. Consider

$$\begin{cases} x' = f(t, x), \\ x(0) = 0, \end{cases}$$

where $f(t, x) : [0, 1] \times \mathbb{R} \to \mathbb{R}$ is given by

$$f(t, x) = \begin{cases} 0, & \text{for } t = 0, \ x \in \mathbb{R}, \\ 2t, & 0 < t \le 1, \ -\infty < x < 0, \\ 2t - \dfrac{4x}{t}, & 0 < t \le 1, \ 0 \le x \le t^2, \\ -2t, & 0 < t \le 1, \ t^2 < x < \infty. \end{cases}$$

All the pieces are continuous except possibly at $t = 0$. The pieces fit together at each break for x. Now near $t = 0$, $\lim_{t \to 0} f(t, x) = 0 = f(0, x)$ in each region implies that f is continuous at $t = 0$ also.

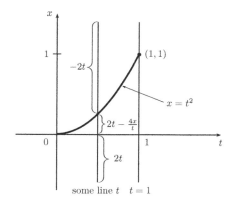

Fig. 1.9   The function $f(t, x)$.

Therefore, f is continuous on $[0, 1] \times \mathbb{R}$. Notice that for a fixed value of $t \in [0, 1]$, $f(t, x)$ is a nonincreasing function of x.

Now let's look at the Picard iterates. Let $x_0(t) = 0$ and $x_n(t) = \int_0^t f(s, x_{n-1}(s))\, ds$. Then,

$$x_1(t) = \int_0^t f(s, 0)\, ds = \int_0^t 2s\, ds = t^2,$$

$$x_2(t) = \int_0^t f(s, s^2)\, ds = \int_0^t \left(2s - \frac{4s^2}{s}\right) ds = \int_0^t -2s\, ds = -t^2,$$

$$x_3(t) = \int_0^t f(s, -s^2)\, ds = \int_0^t 2s\, ds = t^2,$$

$$\vdots$$

$$x_{2n}(t) = -t^2,$$

$$x_{2n+1}(t) = t^2.$$

Therefore, the Picard iterates do not converge to a solution.

Before one last theorem in this section, we will develop a computational aid:

Suppose $f(t, x)$ is continuous on a slab $[t_0, t_0 + \alpha] \times \mathbb{R}^n$ and suppose we have an initial value problem $x' = f(t, x)$, $x(t_0) = x_0$.

Suppose, moreover, that $x(t)$ and $y(t)$ are both solutions of the IVP on $[t_0, t_0 + \alpha]$.

Let the inner product of two vectors $\alpha, \beta \in \mathbb{R}^n$ be denoted

$$\langle \alpha, \beta \rangle = \sum_{i=1}^n \alpha_i \beta_i.$$

Then, on $[t_0, t_0 + \alpha]$,

$$h(t) \equiv \langle x(t) - y(t),\, x(t) - y(t) \rangle = \sum_{i=1}^n (x_i(t) - y_i(t))^2 \geq 0.$$

Note: $h(t_0) = \langle x(t_0) - y(t_0),\, x(t_0) - y(t_0) \rangle = \langle x_0 - x_0,\, x_0 - x_0 \rangle = 0$. Hence, $h(t)$ satisfies the IVP

$$\begin{cases} h'(t) = \sum_{i=1}^n 2(x_i(t) - y_i(t))\, (x_i'(t) - y_i'(t)) \\ \qquad = 2\langle x(t) - y(t),\, x'(t) - y'(t) \rangle, \\ h(t_0) = 0. \end{cases}$$

If $h'(t) \leq 0$ on $[t_0, t_0 + \alpha]$, it follows that $h(t) \equiv 0$ on $[t_0, t_0 + \alpha]$. Thus, $x(t) \equiv y(t)$, if $h'(t) \leq 0$ on $[t_0, t_0 + \alpha]$.

Theorem 1.5. *Let $f(t, x)$ be continuous on the set $Q = \{(t, x)\,|\, t_0 \leq t \leq t_0 + \alpha, \|x - x_0\| \leq b\}$. Assume that, for any $(t, x_1), (t, x_2) \in Q$,*

$$\langle x_1 - x_2,\, f(t, x_1) - f(t, x_2) \rangle \leq 0.$$

Then, if $x(t)$ and $y(t)$ are solutions of the IVP $x' = f(t, x)$, $x(t_0) = x_0$, each with its graph contained in Q for $t_0 \leq t \leq t_0 + \alpha$, it follows that $x(t) \equiv y(t)$ on $[t_0, t_0 + \alpha]$.

Proof. Since $\{(t, x(t))\}, \{(t, y(t))\} \subseteq Q$ and since $x(t)$ and $y(t)$ are solutions of the IVP, we have

$$\langle x(t) - y(t), f(t, x(t)) - f(t, y(t)) \rangle = \langle x(t) - y(t), x'(t) - y'(t) \rangle \leq 0.$$

The middle expression is precisely $\frac{h'(t)}{2}$ from above. Moreover, $h(t_0) = 0$ and so $x(t) \equiv y(t)$ on $[t_0, t_0 + \alpha]$.

Note: When $x' = f(t, x)$ is a scalar equation, (i.e., $f(t, x) : \mathbb{R} \times \mathbb{R} \to \mathbb{R}$), the condition given in the theorem becomes $(x_1 - x_2)(f(t, x_1) - f(t, x_2)) \leq 0$ which is equivalent to saying that $f(t, x)$ is nonincreasing in x, for all fixed t. $\square$

$\boxed{\text{Exercise}}$ **9.** In the preceding Example 1.7, find the unique solution of the IVP on $[0, 1]$. Tell why the solution is unique.

$\boxed{\text{Exercise}}$ **10.** State and prove a theorem corresponding to Theorem 1.5 on $[t_0 - \alpha, t_0]$.

Chapter 2

Continuation of Solutions and Maximal Intervals of Existence

2.1 Continuation of Solutions

Definition 2.1. Let $x(t)$ be a solution of $x' = f(t, x)$ on an interval I. Then a solution $y(t)$ is said to be *a continuation* (or *extension*) of $x(t)$ in case $y(t)$ is a solution on an interval J such that $I \subsetneq J$ and $x(t) \equiv y(t)$ on I.

Theorem 2.1. *Let $f(t, x)$ be continuous on $D \subseteq \mathbb{R} \times \mathbb{R}^n$ and assume that $x(t)$ is a solution of $x' = f(t, x)$ on (a, b) and there is a sequence of t-values with $t_n \uparrow b$ (i.e., $\{t_n\}_{n=1}^{\infty}$ is an increasing sequence and converges to b as $n \to \infty$) such that $\lim_{n \to \infty} x(t_n) = x_0 \in \mathbb{R}^n$. Then, if there exist $\alpha, \beta, M > 0$ such that $\|f(t, x)\| \leq M$ on $D \cap \{(t, x) \mid 0 \leq b - t \leq \alpha, \|x - x_0\| \leq \beta\}$, it follows that $\lim_{t \to b} x(t)$ exists and is equal to x_0. Furthermore, if $f(t, x)$ is continuous on $D \cup \{(b, x_0)\}$, then $x(t)$ can be extended to $(a, b]$ by defining $x(b) = x_0$.*

Note: Theorem 2.1 says the graph of $x(t)$ depicted in Figure 2.1 cannot oscillate in and out of the box for t near b, but must be squeezed into the box; see Figure 2.1.

Proof. We first show that graph of the solution $x(t)$ stays in the box for t near b. Assume the hypotheses of the theorem are satisfied, but that there are values of t arbitrarily close to b such that $\|x(t) - x_0\| > \beta$. Choose m large enough such that $0 < b - t_m < \alpha$ and $\|x(t_m) - x_0\| < \frac{1}{2}\beta$ (we can do this since $t_n \uparrow b$ and $x(t_n) \to x_0$ as $n \to \infty$), and such that $0 < b - t_m < \frac{\beta}{2M}$. By continuity, there exists $t_m < \tau < b$ such that $\|x(\tau) - x_0\| = \beta$ and $\|x(t) - x_0\| < \beta$, for $t_m \leq t < \tau$.

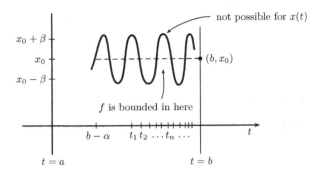

Fig. 2.1 $x(t)$ cannot oscillate in and out of the box for t near b.

Now

$$x(\tau) = x(t_m) + \int_{t_m}^{\tau} f(s, x(s)) \, ds.$$

So,

$$\frac{\beta}{2} = \beta - \frac{\beta}{2} < \|x(\tau) - x_0\| - \|x(t_m) - x_0\| \le \|x(\tau) - x(t_m)\|$$

$$= \left\| \int_{t_m}^{\tau} f(s, x(s)) \, ds \right\| \le \int_{t_m}^{\tau} \|f(s, x(s))\| \, ds$$

$$\le M|\tau - t_m| < M(b - t_m)$$

$$< M \frac{\beta}{2M} = \frac{\beta}{2}.$$

This is a contradiction. Hence, it follows that $\|x(t) - x_0\| < \beta$, for $t_m \le t \le b$. Therefore, the graph of $x(t)$ stays in the box. Then, from

$$x(\sigma) = x(\tau) + \int_{\tau}^{\sigma} f(s, x(s)) \, ds, \quad \forall \sigma, \tau \in (t_m, b),$$

we have

$$\|x(\sigma) - x(\tau)\| \le \left| \int_{\tau}^{\sigma} \|f(s, x(s))\| \, ds \right| \le M|\sigma - \tau|.$$

Consequently, by the Cauchy Criterion $\lim_{t \to b^-} x(t)$ exists. (The Cauchy Criterion is as follows: $\lim_{t \to b^-} x(t)$ exists if and only if for all $\varepsilon > 0$, there exists a $\delta > 0$ such that $\|x(t_1) - x(t_2)\| < \varepsilon$, for any $t_1, t_2 \in (b - \delta, b)$.) But $x(t)$ is continuous and hence $\lim_{t \to b^-} x(t) = x_0$.

Thus, we extend $x(t)$ to $(a, b]$ by defining $x(b) := x_0$. Then $x(t)$ is continuous on $(a, b]$. Moreover, if $f(t, x)$ can be defined at (b, x_0) so as to be continuous on $D \cup \{(b, x_0)\}$, then

$$x'(b) = \lim_{t \to b^-} f(t, x(t)) = f(b, x(b)) = f(b, x_0).$$

Therefore, $x(t)$ has a left-hand derivative at $t = b$ which has the value $f(b, x_0)$. This is a continuation of $x(t)$ to $(a, b]$.   □

Definition 2.2. Let $x(t)$ be a solution of $x' = f(t, x)$ on an interval I. Then the interval I is said to be a *right maximal interval of existence* for $x(t)$ in case there is no proper extension to the right. A *left maximal interval* is defined similarly. I is called a *maximal interval of existence* for $x(t)$ in case it is both right and left maximal.

Definition 2.3. Let $f(t, x)$ be continuous on $D \subseteq \mathbb{R} \times \mathbb{R}^n$ and let $x(t)$ be a solution of $x' = f(t, x)$ on an interval (a, b). Then $x(t)$ is said to *approach the ∂D as $t \to b$*, written $x(t) \to \partial D$ as $t \to b$, in case, for any compact set $K \subseteq D$, there exists $t_K \in (a, b)$ such that $(t, x(t)) \notin K$, for $t_K < t < b$.

Note: If $x(t)$ is a solution on an infinite interval, say $(a, +\infty)$, then $x(t) \to \partial D$ as $t \to \infty$.

Theorem 2.2 (Continuation Theorem). *Let $f(t, x)$ be continuous on an open set $D \subseteq \mathbb{R} \times \mathbb{R}^n$ and let $x(t)$ be a solution of $x' = f(t, x)$ on an interval I. Then $x(t)$ can be continued to be defined on a maximal interval (α, ω) and $x(t) \to \partial D$, as $t \to \alpha$ and as $t \to \omega$.*

Proof. We will prove that $x(t)$ can be continued to be defined on a right maximal interval. Then that extension can be continued to the left to be defined on a left maximal interval.

Let $\{K_n\}_{n=1}^{\infty}$ be a sequence of nonnull open subsets of D such that $\overline{K}_n$ are compact, $\overline{K}_n \subseteq K_{n+1}$, for all $n \geq 1$, and $D = \bigcup_{n=1}^{\infty} K_n$.

$\boxed{\text{Exercise}}$ **11.** Describe how such a sequence of sets can be constructed.

Let b be the right end point of I. If $b = +\infty$, then I is already right maximal. Then as noted above, $x(t) \to \partial D$, as $t \to \infty$.

Second, assume that $b < \infty$ and I is open at b. Then there are two cases:

(1) for all $m \geq 1$, there exists $\tau_m \in I$ such that $(t, x(t)) \notin \overline{K}_m$, for $\tau_m < t < b$, or

(2) there exists a sequence $t_n \uparrow b$ and a set K_m such that $(t_n, x(t_n)) \in \overline{K}_m$, for all $n \geq 1$,

Case 1: We claim if Case 2 holds, then I is already right maximal. So assume Case 2, but suppose I is not right maximal. Then, there exists a proper extension to the right. In particular, if $\tau \in I$, then $x(t)$ is a solution on $[\tau, b]$. Now $\{(t, x(t)) | \tau \leq t \leq b\}$(compact) $\subseteq \bigcup_{n=1}^{\infty} K_n$ which is an open covering. Hence there exists a finite subcovering, so by construction, there exists m_0 such that $\{t, x(t)) | \tau \leq t \leq b\} \subseteq K_{m_0} \subset \overline{K}_{m_0}$, which contradicts to our assumption in Case 2. Thus, in this case, I is right maximal. We note that $x(t) \to \partial D$ as $t \to b$ is also satisfied.

Case 2: Assume that Case 1 holds. Then there exists a sequence $t_n \uparrow b$ and $m \geq 1$ such that $(t_n, x(t_n)) \in \overline{K}_m$ for any $n \geq 1$. Since $\overline{K}_m$ is compact, there exists a subsequence $(t_{n_j}, x(t_{n_j})) \to (b, x_0)$, for some $(b, x_0) \in \overline{K}_m$. By Theorem 2.1, there exists an extension of $x(t)$ to $I \cup \{b\}$.

[From this point on, we will also resolve the case where the interval I is closed at b.]

Now we have $(b, x(b)) \in \overline{K}_m$. By Corollary 1.3 of Theorem 1.4, there exists a $\delta_m > 0$ such that $x(t)$ can be extended to the interval $[b, b + \delta_m]$. If $(b + \delta_m, x(b + \delta_m)) \in \overline{K}_m$, again by Corollary 1.3 of Theorem 1.4, we can extend $x(t)$ to $[b + \delta_m, b + 2\delta_m]$. Continue in this manner. But $\overline{K}_m$ is compact, hence there exists a bound on the number of times this "extending" can be done; i.e., there exists $j_m \geq 1$ such that $(b + j_m \delta_m, x(b + j_m \delta_m)) \notin \overline{K}_m$, but all such previously constructed coordinates belong to $\overline{K}_m$ (i.e., $(b + i\delta_m, x(b + i\delta_m)) \in \overline{K}_m$, $0 \leq i \leq j_m - 1$).

Let $b_1 = b + j_m \delta_m$. It is true that $(b_1, x(b_1)) \in D$. Thus, since $D = \bigcup K_n$, there exists $m_1 > m$ such that $(b_1, x(b_1)) \in \overline{K}_{m_1}$. By the same corollary, there exists a $\delta_{m_1} > 0$ such that $x(t)$ can be extended to $[b_1, b_1 + \delta_{m_1}]$. Repeat the argument above. So, we can say there exists an integer $j_{m_1} \geq 1$ such that $(b_1 + j_{m_1}\delta_{m_1}, x(b_1 + j_{m_1}\delta_{m_1})) \notin \overline{K}_{m_1}$, but $(b_1 + i\delta_{m_1}, x(b + i\delta_{m_1})) \in \overline{K}_{m_1}$, $0 \leq i \leq j_{m_1} - 1$.

As above, let $b_2 = b_1 + j_{m_1}\delta_{m_1}$. Continuing the pattern, we obtain an infinite sequence $b < b_1 < b_2 < b_3 < \cdots$, and an infinite sequence of integers, $m < m_1 < m_2 < \cdots$, such that $x(t)$ is extended to the closed interval $[b, b_j]$ for any $j \geq 1$. Now

$$(b_1, x(b_1)) \notin \overline{K}_m \text{ and } (b_r, x(b_r)) \notin \overline{K}_{m_{r-1}}, r \geq 2. \qquad (2.1)$$

Let $\omega = \sup_{j \geq 1}\{b_j\}$. Then $x(t)$ has been extended to an interval $J = I \cup [b, \omega)$.

Now we establish that J is right maximal. If not, then $x(t)$ has an extension such that $x(t)$ is a solution on $[b, \omega]$. Consider $\{(t, x(t)) \mid b \leq t \leq \omega\}$ (compact) $\subseteq \bigcup K_n$ which is an open covering. Hence, there exists a finite subcover and by the construction of $\{K_n\}$, there exists p such that $\{(t, x(t)) \mid b \leq t \leq \omega\} \subseteq K_p$; a contradiction to (2.1) (i.e., this is a contradiction, since there exists $r - 1 > p$, such that $\overline{K}_p \subseteq \overline{K}_{m_{r-1}}$. Since $b_r < \omega$, we have here that $(b_r, x(b_r)) \in K_p$, yet from (2.1), $(b_r, x(b_r)) \notin \overline{K}_{m_{r-1}}$).

Therefore, $J = I \cup [b, \omega)$ is right maximal. By making a similar argument, $x(t)$ can be extended to the left, so that $x(t)$ can be continued to a maximal interval (α, ω).

Exercise **12.** If $I = (a, \omega)$ and I is right maximal for a solution $x(t)$ then $x(t) \to \partial D$, as $t \to \omega$. Hint: Modify the proof of Cases 1 and 2.

This completes the proof of this theorem. $\square$

If $D = [a, b] \times \mathbb{R}^n$ and $f(t, x)$ is continuous on the slab D, then $f(t, x)$ can be extended to the entire $(n + 1)$-space, $\mathbb{R} \times \mathbb{R}^n$, by

$$f(t, x) = \begin{cases} f(b, x), & t > b, \\ f(a, x), & t < a, \end{cases}$$

and this extension is continuous.

Corollary 2.1. *Let $f(t, x)$ be continuous on $[a, b] \times \mathbb{R}^n$. Then for any $x_0 \in \mathbb{R}^n$, the IVP $x' = f(t, x)$, $x(a) = x_0$ has a solution $x(t)$ which can be continued to be defined on a maximal interval which will be either*

(1) $[a, b]$, *or*
(2) $[a, \omega)$, *where $a < \omega \leq b$ and $\|x(t)\| \to \infty$, as $t \to \omega$.*

Proof. Since f can be extended continuously to $\mathbb{R} \times \mathbb{R}^n$, then by Theorem 2.1, $x(t)$ can be continued to a maximal interval, which is either (1) $[a, b]$, or a proper subinterval (2) $[a, \omega)$, $\omega \leq b$. For 2), define the compact set $K = [a, b] \times \{x \mid \|x\| \leq M\} \subseteq \mathbb{R} \times \mathbb{R}^n \equiv D$. By Theorem 2.1, $x(t) \to \partial D$ as $t \to \omega$. Hence, the graph of $(t, x(t))$ must leave the compact set K. Thus, there exists t_k, $a < t_k < \omega$, such that $(t, x(t)) \notin K$, for $t_k < t < \omega$.

Since $\omega \leq b$, it must be the case that $\|x(t)\| \geq M$, for $t_k < t < \omega$. By the construction of K, $\lim_{t \to \omega^-} \|x(t)\| = +\infty$. As shown in the Figure 2.2, $x(t)$ leaves K at the top or bottom. $\square$

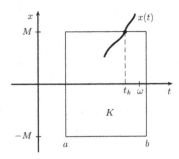

Fig. 2.2 $x(t)$ leaves K at the top.

Example 2.1. Let $f(t,x) = \frac{x}{3t} : D = (0, +\infty) \times \mathbb{R} \to \mathbb{R}$ and consider

$$\begin{cases} x' = f(t,x), \\ x(1) = 1. \end{cases}$$

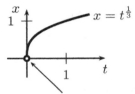

The line approaches the vertical!

Fig. 2.3 $x(t) \to \partial D$.

By separation of variables, the solution is $x = t^{\frac{1}{3}}$. The interval $(0, +\infty)$ is left-maximal and notice that, given any compact $K \subset (0, +\infty) \times \mathbb{R}$, there exists $t_K > 0$ such that $(t, t^{\frac{1}{3}}) \notin K$, $\forall 0 < t < t_K$. Hence, $x(t) \to \partial D$ as $t \to 0$. Moreover, given any compact $K \subset (0, +\infty) \times \mathbb{R}$, there exists $t_K > 0$ $(t, t^{\frac{1}{3}}) \notin K$ for any $t_K < t < +\infty$. Hence, $x(t) \to \partial D$ as $t \to \infty$. See Figure 2.3.

Corollary 2.2. *Let $f(t,x)$ be continuous on the set $Q = \{(t,x) \mid |t - t_0| \leq a, \|x - x_0\| \leq b\}$. Let $M = \max_{(t,x) \in Q} \|f(t,x)\|$. Then all solutions of the IVP $x' = f(t,x)$, $x(t_0) = x_0$ extend to the interval $[t_0 - \alpha, t_0 + \alpha]$ where $\alpha = \min\{a, \frac{b}{M}\}$.*

Proof. First, we extend f continuously to $\mathbb{R} \times \mathbb{R}^n$ in the following way. Let t, with $|t - t_0| \leq a$, be fixed and consult the Figure 2.4.

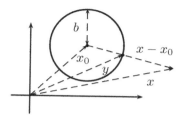

Fig. 2.4   The ball centered at x_0 with radius b.

By vector addition, $y = x_0 + \lambda(x - x_0)$, where $\lambda > 0$. Since y is on the ball, $\|y - x_0\| = b$. So, $\|x_0 + \lambda(x - x_0) - x_0\| = b$, that is, $\lambda = \frac{b}{\|x - x_0\|}$. Thus, for any x outside the ball of radius b, there exists y on the boundary of the ball such that $y = x_0 + \frac{b}{\|x - x_0\|}(x - x_0)$, which pulls x down to the ball. Thus for x outside the ball, define

$$f(t, x) = f(t, y), \qquad \text{where } y = x_0 + \frac{b}{\|x - x_0\|}(x - x_0).$$

In this way f is extended continuously to $[t_0 - a, \ t_0 + a] \times \mathbb{R}^n$.

Now let t vary and consult the Figure 2.5.

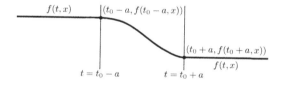

Fig. 2.5   Extension of $f(t, x)$ in t.

We extend f further by defining

$$f(t, x) = \begin{cases} f(t_0 - a, x), & \text{if } t < t_0 - a, \\ f(t, x), & \text{if } t_0 - a \leq t \leq t_0 + a, \\ f(t_0 + a, x), & \text{if } t > t_0 + a. \end{cases}$$

Again f is continuous and thus f is extended continuously to $\mathbb{R} \times \mathbb{R}^n$.

Now let $x(t)$ be a solution of the IVP $x' = f(t, x)$, $x(t_0) = x_0$, where $f(t, x)$ is the extension above. We will argue the assertion of the theorem to the right; i.e., that $x(t)$ can be extended as a solution to $[t_0, t_0 + \alpha]$.

Let $[t_0, \omega)$ be a right maximal interval of existence for $x(t)$. Since Q is compact, there exists t_Q such that $(t, x(t)) \notin Q$ for $t_Q < t < \omega$. By continuity, there exists some t_1 such that $t_0 \leq t_1 \leq t_Q$, $(t_1, x(t_1)) \in \partial Q$, and $(t, x(t)) \in \text{Int } Q$, for $t_0 \leq t \leq t_1$.

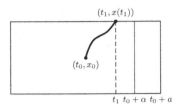

Fig. 2.6   The graph of $(t, x(t))$ hits ∂Q at t_1.

Assume that $t_0 < t_1 < t_0 + \alpha$ (i.e., assume the graph of $(t, x(t))$ hits ∂Q before t reaches $t_0 + \alpha$) (Note: $t_1 - t_0 < \alpha$). Then $\|x(t_1) - x_0\| = b$. Since Q is closed, $(t, x(t)) \in Q$, $t_0 \leq t \leq t_1$. See Figure 2.6. Therefore, $\|f(t, x(t))\| \leq M$, for $t_0 \leq t \leq t_1$. Now from $x(t_1) = x_0 + \int_{t_0}^{t_1} f(s, x(s))\, ds$, we have

$$b = \|x(t_1) - x_0\| \leq \int_{t_0}^{t_1} \|f(s, x(s))\|\, ds \leq M(t_1 - t_0) < M\alpha \leq b,$$

which is a contradiction. Hence, our assumption that $t_0 < t_1 < t_0 + \alpha$ is false, which implies that $t_1 \geq t_0 + \alpha$, and consequently $x(t)$ extends to $[t_0, t_0 + \alpha]$.

Since $x(t)$ was an arbitrary solution of the IVP, it follows that every solution extends to $[t_0, t_0 + \alpha]$. Similar arguments can be made for extending to $[t_0 - \alpha, t_0]$.   □

Corollary 2.3. *Let $f(t, x)$ be continuous and satisfy a local Lipschitz condition wrt x on the open set $D \subseteq \mathbb{R} \times \mathbb{R}^n$. Then solutions of IVP's for $x' = f(t, x)$ are globally unique in D.*

Proof. Let $(t_0, x_0) \in D$, $x(t)$ and $y(t)$ be solutions of $x' = f(t, x)$, $x(t_0) = x_0$, with respective maximal intervals of existence (α_1, w_1) and (α_2, w_2). We prove the solutions are unique to the right.

Let $\tau = \sup \{s \geq t_0 \mid x(t) \equiv y(t) \text{ on } [t_0, s]\}$. Then $x(t) \equiv y(t)$ on $[t_0, \tau)$.

Assume that $t_0 \leq \tau < \min \{w_1, w_2\}$. Then both $x(t)$ and $y(t)$ are continuous on $[t_0, \tau]$, and thus by this continuity, $x(t) \equiv y(t)$ on $[t_0, \tau]$.

By the Picard Theorem, there exists $\alpha > 0$ such that solutions of the IVP

$$\begin{cases} z' = f(t, z), \\ z(\tau) = x(\tau) \end{cases}$$

exist and are unique on $[\tau - \alpha, \tau + \alpha]$. Thus, $y(t) \equiv x(t)$ on $[\tau - \alpha, \tau + \alpha]$, hence on $[t_0, \tau + \alpha]$. But this contradicts the definition of τ. So, $\tau < \min\{\omega_1, \omega_2\}$ is false, and hence, $\tau = \min\{\omega_1, \omega_2\}$ (τ cannot be larger, since (α_1, ω_1) and (α_2, ω_2) were maximal intervals).

We claim that $\omega_1 = \omega_2$. If $\omega_1 < \omega_2$, so that $\tau = \omega_1$, then $x(t) \equiv y(t)$ on $[t_0, \omega_1)$, and $y(t)$ would constitute an extension of $x(t)$ to $[t_0, \omega_2)$. But $x(t)$ cannot be extended past ω_1. Thus $\omega_1 \geq \omega_2$. Arguing similarly, $\omega_2 \geq \omega_1$, so that $\omega_1 = \omega_2$.

Therefore, $x(t) \equiv y(t)$ on $[t_0, \omega_1) = [t_0, \omega_2)$. $\square$

Exercise 13. Make the argument in the corollary to the left.

Exercise 14. Let $f(t, x)$ be a continuous real-valued function on $Q = \{(t, x) \in \mathbb{R} \times \mathbb{R} \mid |t - t_0| \leq a, |x - x_0| \leq b\}$. Let $M = \max_{(t,x) \in Q} |f(t, x)|$ and $\alpha = \min\{a, \frac{b}{M}\}$. If $x_1(t), x_2(t), \ldots, x_n(t)$ are solutions of

$$\begin{cases} x' = f(t, x), \\ x(t_0) = x_0 \end{cases} \tag{2.2}$$

on $[t_0 - \alpha, t_0 + \alpha]$, prove that $z(t) = \max_{1 \leq j \leq n} x_j(t)$ and $y(t) = \min_{1 \leq j \leq n} x_j(t)$ are solutions on $[t_0 - \alpha, t_0 + \alpha]$.

Remark 2.1. It is also the case that $\varphi(t) = \sup\{x(t) \mid x(t) \text{ is a solution of } (2.2)\}$ and $\psi(t) = \inf\{x(t) \mid x(t) \text{ is a solution of } (2.2)\}$ are solutions of (2.2) on $[t_0 - \alpha, t_0 + \alpha]$.

2.2   Kamke Convergence Theorem

Theorem 2.3 (Kamke Convergence Theorem). *Let $D \subseteq \mathbb{R} \times \mathbb{R}^m$ be open and let $\{f_n(t, x)\}_{n=1}^{\infty}$ be a sequence of continuous m-vector valued functions on D such that $\lim_{n \to \infty} f_n(t, x) = f_0(t, x)$ uniformly on each compact subset of D. For each $n \geq 1$, let $x_n(t)$ be a noncontinuable solution of the IVP*

$$\begin{cases} x' = f_n(t, x), \\ x(t_n) = y_n, \quad \text{where } (t_n, y_n) \in D. \end{cases} \tag{2.3$_n$}$$

Assume $x_n(t)$ is defined on (α_n, ω_n). Further assume $\lim_{n \to \infty} (t_n, y_n) = (t_0, y_0) \in D$. Then there exists a noncontinuable solution $x_0(t)$ of

$$\begin{cases} x' = f_0(t, x), \\ x(t_0) = y_0, \end{cases} \tag{2.3}_0$$

with interval of existence (α_0, ω_0), and there exists a subsequence $\{x_{n_k}(t)\}$ of the sequence $\{x_n(t)\}$ such that, for each compact $[a, b] \subset (\alpha_0, \omega_0)$, $\lim_{k \to \infty} x_{n_k}(t) = x_0(t)$ uniformly on $[a, b]$, in the sense that, there is an $N_{[a,b]} \in \mathbb{N}$ such that for $n_k \geq N_{[a,b]}$, the compact $[a, b] \subset (\alpha_{n_k}, \omega_{n_k})$ and $\lim_{h \to \infty} x_{n_k}(t) = x_0(t)$ uniformly on $[a, b]$.

$\boxed{\textbf{Exercise}}$ **15.** Prove that, from the last statement of the theorem, $\overline{\lim}_{k \to \infty} \alpha_{n_k} \leq \alpha_0$ and $\omega_0 \leq \underline{\lim}_{k \to \infty} \omega_{n_k}$; in particular $\overline{\lim}_{k \to \infty} \alpha_{n_k} \leq \alpha_0 < \omega_0 \leq \underline{\lim}_{k \to \infty} \omega_{n_k}$.

Proof of the theorem. We will prove that there is a solution $x_0(t)$ of $(2.3)_0$ on a right maximal interval of existence $[t_0, \omega_0)$ and a subsequence $\{x_{n_k}(t)\}$ of the sequence $\{x_n(t)\}$ such that for any τ with $t_0 < \tau < \omega_0$, there is a $N_{[t_0,\tau]}$ such that $[t_0, \tau] \subset (\alpha_{n_k}, \omega_{n_k})$, for $n_k \geq N_{[t_0,\tau]}$ and $\lim_{k \to \infty} x_{n_k}(t) = x_0(t)$ uniformly on $[t_0, \tau]$. See Figure 2.7. Having done this, we return to the original statement of Theorem 2.3 and replace the original sequence $\{x_n(t)\}$ by the subsequence $\{x_{n_k}(t)\}$; then carry out an analogous procedure on a subsequence $\{x_{n_{k_j}}(t)\}$ of $\{x_{n_k}(t)\}$ to get a limit solution on a left maximal interval.

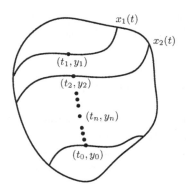

Fig. 2.7   The solution sequence $\{x_n(t)\}$.

Proceeding with the proof, let $\{K_n\}_{n=1}^{\infty}$ be a sequence of nonnull open sets such that $\overline{K}_n$ are compact, $\overline{K}_n \subseteq K_{n+1}$, and $\bigcup K_n = D$. See Figure 2.8.

For each $j \geq 1$, let $\rho_j = \text{dist}(\overline{K}_j$, complement of $K_{j+1}) > 0$. Let $K_j^* = \{(t,x) \mid \text{dist}((t,x),\overline{K}_j) \leq \frac{1}{2}\rho_j\}$. Then K_j^* is compact and $K_j^* \subseteq K_{j+1}$. Now as $n \to \infty$, the sequence $f_n(t,x) \to f_0(t,x)$ on K_j^* uniformly, which implies that there exists $M_j > 0$ such that $\|f_n(t,x)\| \leq M_j$ on K_j^* for all $n \geq 0$. It follows that, for any point $(s,z) \in \overline{K}_j$ and any $n \geq 0$, the IVP

$$\begin{cases} x' = f_n(t,x), \\ x(s) = z, \end{cases}$$

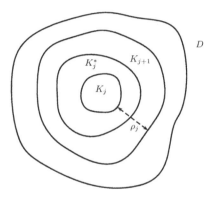

Fig. 2.8 $\overline{K}_n \subseteq K_{n+1}$, and $\bigcup K_n = D$.

has all of its solutions defined on $[s-\delta_j, s+\delta_j]$, where $\delta_j = \min\{\frac{\rho_j}{4}, \frac{\rho_j}{4M_j}\}$ by Corollary 2.2. (Note: The calculation of δ_j is as follows: Take $(s,z) \in \overline{K}_j$ and $Q = \{(t,x) \mid |t-s| \leq \frac{1}{4}\rho_j, \|x-z\| \leq \frac{1}{4}\rho_j\}$. So, $Q \subseteq K_j^*$ and then set $\delta_j = \alpha_j$ as has been done previously in Corollary 2.2.)

Also, all solutions are uniformly bounded on $[s-\delta_j, s+\delta_j]$, since $\|x(t)\| \leq \|x(t) - z\| + \|z\| \leq \frac{1}{4}\rho_j + \sup_{z \in K_j^*}\|z\|$, where $z \in \overline{K}_j$.

Furthermore, all solutions satisfy a Lipschitz condition, since for a typical solution $x(t) = x(s) + \int_s^t f_n(\hat{s}, x(\hat{s}))\, d\hat{s}$ and $x(\tau) = x(s) + \int_s^\tau f_n(\hat{s}, x(\hat{s}))\, d\hat{s}$, where $\tau, t \in [s-\delta_j, s+\delta_j]$. Hence,

$$\|x(t) - x(\tau)\| \leq \left| \int_\tau^t \|f_n(\hat{s}, x(\hat{s}))\|\, d\hat{s} \right| \leq M_j|t - \tau|$$

holds for all $t, \tau \in [s - \delta_j, s + \delta_j]$; that is, *every solution* to each IVP $(2.3)_j$ satisfies the same Lipschitz condition.

Now assume $(t_0, y_0) \in K_{m_1}$. Then there is an N_1 such that, for all $n \geq N_1$, $(t_n, y_n) \in K_{m_1}$ and $|t_n - t_0| < \varepsilon_1 = \frac{1}{2}\delta_{m_1}$, since $(t_n, y_n) \to (t_0, y_0)$. See Figure 2.9.

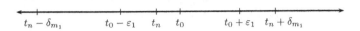

Fig. 2.9   The points t_0 and t_n.

At this stage we have, for $n \geq N_1$, $[t_0, t_0 + \varepsilon_1] \subseteq [t_n - \delta_{m_1}, t_n + \delta_{m_1}] \subset (\alpha_n, \omega_n)$, because all solutions of the IVP exist on $[t_n - \delta_{m_1}, t_n + \delta_{m_1}]$. Furthermore, $\{x_n(t)\}_{n=N_1}^{\infty}$ is uniformly bounded and equicontinuous on $[t_0, t_0 + \varepsilon_1]$. Therefore, by the Arzelà-Ascoli Theorem, there is a subsequence of integers $\{n_{1(k)}\}_{k=1}^{\infty} \subseteq \{n\}$ such that the subsequence $\{x_{n_{1(k)}}(t)\}$ converges uniformly on $[t_0, t_0 + \varepsilon_1]$. Call this limit $x_0(t)$; i.e., $\lim_{k \to \infty} x_{n_{1(k)}}(t) := x_0(t)$ on $[t_0, t_0 + \varepsilon_1]$ uniformly.

We **claim** that $x_0(t)$ is a solution of $x' = f_0(t, x)$, $x(t_0) = y_0$ on $[t_0, t_0 + \varepsilon_1]$.

Let $t \in [t_0, t_0 + \varepsilon_1]$. Then

$$x_{n_{1(k)}}(t)$$
$$= y_{n_{1(k)}} + \int_{t_{n_{1(k)}}}^{t} f_{n_{1(k)}}(s, x_{n_{1(k)}}(s)) \, ds$$
$$= y_{n_{1(k)}} + \int_{t_0}^{t} f_{n_{1(k)}}(s, x_{n_{1(k)}}(s)) \, ds + \int_{t_{n_{1(k)}}}^{t_0} f_{n_{1(k)}}(s, x_{n_{1(k)}}(s)) \, ds,$$

since $[t_0, t_0 + \varepsilon_1] \subseteq [t_{n_{1(k)}} - \delta_{m_1}, t_{n_{1(k)}} + \delta_{m_1}]$.

Let $n_{1(k)} \to \infty$. Then $x_{n_{1(k)}}(t) \to x_0(t)$ and $f_{n_{1(k)}}(s, x_{n_{1(k)}}(s)) \to f_0(s, x_0(s))$, and so,

$$x_0(t) = y_0 + \int_{t_0}^{t} f_0(s, x_0(s)) \, ds, \text{ for } t \in [t_0, t_0 + \varepsilon_1].$$

This verifies the claim and so $x_0(t)$ is a solution of $x' = f_0(t, x)$, $x(t_0) = y_0$ on $[t_0, t_0 + \varepsilon_1]$.

Now, if $(t_0 + \varepsilon_1, x_0(t_0 + \varepsilon_1)) \in K_{m_1}$ (i.e., the point on the graph of $x_0(t)$ at the end point of $[t_0, t_0 + \varepsilon_1]$), then we can repeat this process, since $(t_0 + \varepsilon_1, x_{n_{1(k)}}(t_0 + \varepsilon_1)) \to (t_0 + \varepsilon_1, x_0(t_0 + \varepsilon_1))$ (i.e., the process we have gone through depends only on the fact that $(t_0 + \varepsilon_1, x_0(t_0 + \varepsilon_1)) \in K_{m_1}$). Hence, repeating the above process, we obtain a second subsequence $\{n_{2(k)}\} \subseteq \{n_{1(k)}\}$ such that $\lim_{k\to\infty} x_{n_{2(k)}}(t) \equiv x_0(t)$ uniformly on $[t_0 + \varepsilon_1, t_0 + 2\varepsilon_1]$, and consequently $\lim_{k\to\infty} x_{n_{2(k)}}(t) \equiv x_0(t)$ uniformly on $[t_0, t_0 + 2\varepsilon_1]$.

Continuing in this manner, we must reach a first integer $j_1 \geq 1$ such that $(t_0 + j_1\varepsilon_1,\ x_0(t_0 + j_1\varepsilon_1)) \notin K_{m_1}$, and we will have also obtained corresponding subsequences $\{n_{i(k)}\}$, $1 \leq i \leq j_1$, such that $\{n_{(i+1)(k)}\} \subseteq \{n_{i(k)}\}$. Define $\hat{t}_1 = t_0 + j_1\varepsilon_1$, and assume that $(\hat{t}_1, x_0(\hat{t}_1)) \in K_{m_2}$, for m_2 (recall $D = \bigcup K_n$). Then we start over and repeat our construction as above.

Summarizing, we obtain sequences, $t_0 < \hat{t}_1 < \hat{t}_2 < \cdots$, where $\hat{t}_1 = t_0 + j_1\varepsilon_1$, $\hat{t}_2 =$ the first point where you go outside K_{m_2}, etc, and a sequence of sets $K_{m_1} \subseteq K_{m_2} \subseteq K_{m_3} \subseteq \cdots$, with a subsequence of integers $\{n_{1(k)}\}, \{n_{2(k)}\}, \{n_{3(k)}\}, \ldots$.

Let $\omega_0 = \sup_{j \geq 1}\{\hat{t}_j\}$.

We **claim** that $x_0(t)$ is a solution of $x' = f_0(t, x)$, $x(t_0) = y_0$ on $[t_0, \omega_0)$.

To see that this is a solution, let $t_0 < \tau < \omega_0$. Then by the definition of ω_0, there exists $t_0 < \tau < \hat{t}_p < \omega_0$. By our construction, there is a subsequence $\{x_{n_{i_0(k)}}(t)\}$ which converges uniformly to $x_0(t)$ on $[t_0, \hat{t}_p]$, hence, $x_0(t)$ is a solution on $[t_0, \hat{t}_p]$ and therefore is a solution on $[t_0, \tau]$. But τ was arbitrarily selected, thus $x_0(t)$ is a solution on $[t_0, \omega_0)$. Consider now the array of sequences:

$$\{n_{1(k)}\} = n_{1(1)}\ n_{1(2)}\ n_{1(3)}\ \cdots$$
$$\cup|$$
$$\{n_{2(k)}\} = n_{2(1)}\ n_{2(2)}\ n_{2(3)}\ \cdots$$
$$\cup|$$
$$\{n_{3(k)}\} = n_{3(1)}\ n_{3(2)}\ n_{3(3)}\ \cdots$$
$$\cup|$$
$$\vdots \qquad \vdots \quad \vdots \quad \vdots \quad \cdots$$

Take the diagonal subsequence $\{n_k\} \equiv \{n_{k(k)}\}$. With this subsequence, $\{x_{n_k}\}$ converges uniformly to $x_0(t)$ on each compact subinterval of $[t_0, \omega_0)$. To see this, let $[a, b] \subseteq [t_0, \omega_0)$. Then again, there exists $t_0 < b < \hat{t}_p < \omega_0$ so that, $[a, b] \subseteq [t_0, \hat{t}_p]$.

By our construction, there is a subsequence $\{x_{n_{i_0(k)}}(t)\}$ which converges uniformly to $x_0(t)$ on $[t_0, \hat{t}_p]$, hence on $[a, b]$. However, $\{n_{k(k)}\}_{k \geq i_0} \subseteq \{n_{i_0(k)}\}$. Thus, for $k \geq i_0$, it will follow that $[t_0, \omega_{n_k}] \supseteq [a, b]$ and for $k \geq i_0$, $\lim_{n_k \to \infty} x_{n_k}(t) \equiv x_0(t)$ uniformly on $[a, b]$. Thus, the convergence properties stated in this theorem are satisfied on $[t_0, \omega_0)$.

Moreover, it follows from the fact that $(\hat{t}_p, x_0(\hat{t}_p)) \notin K_{m_p}$, for each p, and that $[t_0, \omega_0)$ is right maximal for $x_0(t)$.

To complete the proof of the theorem, take the sequence $\{x_{n_k}(t)\}$ and let it play the role of the original sequence. Make the analogous argument to the left of t_0. The subsequence constructed this time will be a subsequence of $\{x_{n_k}\}$ which is in turn a subsequence of $\{x_n\}$. $\qquad\square$

Remark 2.2. The Kamke Theorem is valid in the case where D is a slab of the form $D = [a, b] \times \mathbb{R}^m$. Suppose as in the theorem that $f_n \to f_0$ on all compact subsets of D, and that $(t_n, y_n) \to (t_0, y_0)$, etc. See Figure 2.10.

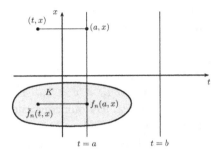

Fig. 2.10   The Kamke Theorem is valid on $D = [a, b] \times \mathbb{R}^m$.

All conditions of the theorem are satisfied by making the extension:

$$\bar{f}_n(t, x) = \begin{cases} f_n(a, x), & \text{if } t < a, \\ f_n(t, x), & \text{if } a \le t \le b, \\ f_n(b, x), & \text{if } t > b. \end{cases}$$

Then if K is a compact set of the open set $\mathbb{R} \times \mathbb{R}^m$, then

$$\sup_{(t,x)\in K} \left\| \bar{f}_n(t, x) - \bar{f}_0(t, x) \right\| = \sup_{(t,x)\in K\cap D} \left\| f_n(t, x) - f_0(t, x) \right\|.$$

In particular, we can satisfy the uniform convergence of $\{\bar{f}_n\}$ to $\bar{f}_0$ on any compact $K \subseteq \mathbb{R} \times \mathbb{R}^m$. Thus, in the presence of the other hypotheses of the theorem, the Kamke Theorem applies equally well for slab regions.

2.3   Continuous Dependence of Solutions on Initial Conditions

In this and next sections, we consider consequences of the Kamke Theorem, with one of the main consequences dealing with the continuous dependence of solutions of IVP's depending on initial conditions and parameters.

Corollary 2.4 (to the Kamke Theorem). *Assume the hypotheses of the Kamke Theorem and in addition that the limit IVP* $(2.3)_0$ *has a unique noncontinuable solution* $x_0(t)$ *with maximal interval* (α_0, ω_0).

Then, given any compact interval $[a, b] \subseteq (\alpha_0, \omega_0)$, *there exists* N *such that*

(i) $[a, b] \subseteq (\alpha_n, \omega_n)$, *for all* $n \geq N$, *and*
(ii) $\lim_{n \to \infty} x_n(t) = x_0(t)$, *for* $n \geq N$, *uniformly on* $[a, b]$.

Proof. Assume that (i) is false; that is, there is no N such that $[a, b] \subseteq (\alpha_n, \omega_n)$, for all $n \geq N$. Then there exists a subsequence $\{x_{n_k}(t)\}$ such that $[a, b] \not\subseteq (\alpha_{n_k}, \omega_{n_k})$. In the Kamke Theorem, replace the original sequence with this subsequence. Relative to this subsequence, the hypotheses of the Kamke Theorem are satisfied. By the Kamke Theorem, there exists a subsequence $\{x_{n_{k_i}}(t)\}$ such that $\{x_{n_{k_i}}(t)\}$ converges to a solution of IVP $(2.3)_0$ with the convergence being uniform on each compact subinterval of the interval of existence for the solution of $(2.3)_0$.

But $x_0(t)$ is the unique solution of $(2.3)_0$, hence $\{x_{n_{k_i}}(t)\}$ converges to $x_0(t)$ uniformly each compact subset of (α_0, ω_0), but this is a contradiction. Thus (i) holds.

Assume now that (ii) is false. Then, there exists $\varepsilon_0 > 0$ and a subsequence $\{x_{n_k}(t)\} \subseteq \{x_n(t)\}_{n \geq N}$ such that $\|x_{n_k}(t) - x_0(t)\| \geq \varepsilon_0$ at some points $t \in [a, b]$ and for all n_k. Now replace the sequence in the Kamke Theorem with this subsequence. By the Kamke Theorem, it follows that there exists a subsequence $\{x_{n_{k_i}}(t)\} \subseteq \{x_{n_k}(t)\}$ which converges uniformly to a solution, to $x_0(t)$ by uniqueness, of $(2.3)_0$ on each compact subinterval of (α_0, ω_0); in particular, for i sufficiently large, we have $\|x_{n_{k_i}}(t) - x_0(t)\| < \varepsilon_0$, for all $t \in [a, b]$, which is a contradiction.

Therefore, $\lim_{n \to \infty} x_n(t) = x_0(t)$ uniformly on $[a, b]$ for $n \geq N$.   □

Remark 2.3. Above, it may be the case that the IVP's

$$\begin{cases} x' = f_n(t, x), \\ x(t_n) = y_n, \end{cases} \qquad (2.3)_n$$

do not have unique solutions. Yet, if $x_0(t)$ is unique, then any sequence of solutions of $(2.3)_n$ converge to $x_0(t)$ uniformly on compact subsets of (α_0, ω_0).

For our next consequence of the Kamke Theorem, we see that if solutions of IVP's are unique, then whenever initial conditions are perturbed slightly, the resulting solutions stay uniformly near to each other.

Theorem 2.4. (Continuous Dependence of Solutions on Initial Conditions) *Let $f(t,x)$ be continuous on an open set $D \subseteq \mathbb{R} \times \mathbb{R}^n$ (also holds on slabs with the appropriate extensions), and assume that IVP's for $x' = f(t,x)$ on D have unique solutions. Given any $(t_0, x_0) \in D$, let $x(t; t_0, x_0)$ denote the solution of*

$$\begin{cases} x' = f(t,x), \\ x(t_0) = x_0 \end{cases}$$

with maximal interval $(\alpha(t_0, x_0), \omega(t_0, x_0))$. Then for each $\varepsilon > 0$ and for each compact $[a,b] \subseteq (\alpha(t_0, x_0), \omega(t_0, x_0))$, there exists a $\delta > 0$ such that for all $(t_1, x_1) \in D$, $|t_0 - t_1| < \delta$ and $\|x_1 - x_0\| < \delta$ imply that $[a,b] \subseteq (\alpha(t_1, x_1), \omega(t_1, x_1))$, the maximal interval of existence of the solution $x(t; t_1, x_1)$ of

$$\begin{cases} x' = f(t,x), \\ x(t_1) = x_1, \end{cases}$$

and $\|x(t; t_1, x_1) - x(t; t_0, x_0)\| < \varepsilon$ on $[a,b]$.

Proof. Assume that there is a $(t_0, x_0) \in D$, such that $[a,b] \subseteq (\alpha(t_0, x_0), \omega(t_0, x_0))$, and an $\varepsilon > 0$ such that no such δ exists. See Figure 2.11. Choose a sequence $\{\delta_n\} \downarrow 0$ such that one or the other of the conclusions fail for each δ_n. Hence, for each n, there exists $(t_n, x_n) \in D$ with $|t_0 - t_n| < \delta_n$, $\|x_0 - x_n\| < \delta_n$. Then, $(t_n, x_n) \to (t_0, x_0)$, since $\delta_n \downarrow 0$, with one or the other of the conclusions failing for the corresponding solution $x(t; t_n, x_n)$ of the IVP

$$\begin{cases} x' = f(t,x), \\ x(t_n) = x_n. \end{cases}$$

But this violates the previous corollary, since this sequence $\{x(t; t_n, x_n\}$ satisfies the hypotheses of the Kamke Theorem, plus we have a unique solution $x(t; t_0, x_0)$.

Thus, our assumption is false and the conclusions must hold. $\qquad\square$

$\boxed{\text{Exercise}}$ **16.** With the hypotheses and notations of Theorem 2.4, prove that $\alpha(t,x)$ is upper semi-continuous and $\omega(t,x)$ is lower semi-continuous on D.

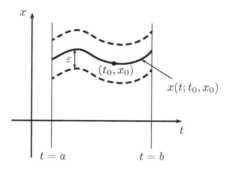

Fig. 2.11   A solution $x(t; t_0, x_0)$.

Remark 2.4. Suppose the slab $D = [a, b] \times \mathbb{R}^n$ is given and that $f(t, x)$ is continuous on D and that solutions of IVP's are unique on D. Assume that $x_0(t)$ is a solution of $x' = f(t, x)$ which exists on $[a, b]$. Then any solution within ε of $x_0(t)$ will exist also on $[a, b]$ by the Kamke Theorem and Theorem 2.4.

2.4   Continuity of Solutions wrt Parameters

Another type of continuous dependence satisfied by solutions, in the presence of certain uniqueness conditions, is the **continuity of solutions wrt parameters**. For example, we might consider the dependence of solutions upon the parameter λ in equations of the form

$$(1 - t^2)x'' - 2tx' + \lambda(\lambda + 1)x = 0 \quad \text{(Legendre's Equation)},$$

or

$$t^2 x'' + tx' + (t^2 - \lambda^2)x = 0 \quad \text{(Bessel's Equation of order } \lambda).$$

Theorem 2.5 (Continuity of Solutions wrt Parameters). *Let $f(t, x, \lambda)$ be continuous on $D_1 \times D_2$, where $D_1 \subseteq \mathbb{R} \times \mathbb{R}^n$ and $D_2 \subseteq \mathbb{R}^m$ are open.*

Assume that for each $(t_0, x_0, \lambda_0) \in D \equiv D_1 \times D_2$, the solution of the IVP

$$\begin{cases} x' = f(t, x, \lambda_0), \\ x(t_0) = x_0 \end{cases}$$

is unique. Then, for each $(t_0, x_0, \lambda_0) \in D$ and for each $[a, b] \subset (\alpha(t_0, x_0, \lambda_0), \omega(t_0, x_0, \lambda_0))$ and for each $\varepsilon > 0$, there exists $\delta > 0$ such that

$(t_0, x_0, \lambda_1) \in D$ and $\|\lambda_1 - \lambda_0\| < \delta$ imply $[a, b] \subset (\alpha(t_0, x_0, \lambda_1), \omega(t_0, x_0, \lambda_1))$ and that $\|x(t; t_0, x_0, \lambda_1) - x(t; t_0, x_0, \lambda_0)\| < \varepsilon$ on $[a, b]$.

Note: We are denoting the solution of the IVP $x' = f(t, x, \lambda_0)$, $x(t_0) = x_0$ by $x(t; t_0, x_0, \lambda_0)$.

Proof. Set

$$z = \begin{bmatrix} x \\ \lambda \end{bmatrix},$$

i.e., z is an $n + m$ vector with the first n components those of x and the last m components those of λ.

Set

$$h(t, z) = \left.\begin{bmatrix} f_1(t, x, \lambda) \\ f_2(t, x, \lambda) \\ \vdots \\ f_n(t, x, \lambda) \\ 0 \\ \vdots \\ 0 \end{bmatrix}\begin{array}{l} \left.\vphantom{\begin{matrix} f_1 \\ f_2 \\ \vdots \\ f_n \end{matrix}}\right\} n \\ \\ \left.\vphantom{\begin{matrix} 0 \\ \vdots \\ 0 \end{matrix}}\right\} m \end{array}\right. = \begin{bmatrix} h_1(t, z) \\ h_2(t, z) \\ \\ \vdots \\ \\ h_{n+m}(t, z) \end{bmatrix},$$

and the consider the IVP

$$\begin{cases} z' = h(t, z), \\ z(t_0) = z_0 = \begin{bmatrix} x(t_0) \\ \lambda_0 \end{bmatrix} = \begin{bmatrix} x_0 \\ \lambda_0 \end{bmatrix}, \end{cases}$$

which is now an IVP without parameter. We note that $z_i' = 0$, for $n + 1 \le i \le n + m$, thus that in the solution, $z(t)$, the components $z_i(t)$ are constant, for $n + 1 \le i \le n + m$. From the initial condition, we must have $z_{n+i}(t_0) = \lambda_{0_i}$, $1 \le i \le m$; thus $z_{n+i}(t) \equiv \lambda_{0_i}$, $1 \le i \le m$.

Hence, the λ's in $f_1(t, x, \lambda)$, $f_2(t, x, \lambda)$, ..., are these λ_{0_i}'s, i.e., $z_i(t) = x_i(t)$, $1 \le i \le n$, are the components of the solution of $x' = f(t, x, \lambda_0)$, $x(t_0) = x_0$.

Therefore, the problem

$$\begin{cases} x' = f(t, x, \lambda_0), \\ x(t_0) = x_0 \end{cases} \qquad \text{is equivalent to} \qquad \begin{cases} z' = h(t, z), \\ z(t_0) = \begin{bmatrix} x_0 \\ \lambda_0 \end{bmatrix}. \end{cases}$$

Apply Theorem 2.4 to $z' = h(t, z)$ (note that the solutions $z_i(t)$ are unique since solutions to $x_i' = f_i(t, x, \lambda)$ are unique), and the conclusions follow. $\square$

Remark 2.5. Suppose that $f(t, x)$ is continuous on a slab region of the form $[t_0, \infty) \times \mathbb{R}^n$. Further assume that the solution of the IVP

$$\begin{cases} x' = f(t, x), \\ x(t_0) = x_0, \end{cases}$$

is unique and exists on $[t_0, \infty)$. Then by our results concerning continuity of solutions wrt the initial conditions, given $[t_0, t_1]$ and given $\varepsilon > 0$, there exists $\delta > 0$ such that $\|x_1 - x_0\| < \delta$ implies all solutions of

$$\begin{cases} x' = f(t, x), \\ x(t_0) = x_1 \end{cases}$$

extend to $[t_0, t_1]$ and satisfy $\|x(t; t_0, x_1) - x(t; t_0, x_0)\| < \varepsilon$ on $[t_0, t_1]$.

Definition 2.4. The solution $x(t; t_0, x_0)$ is said to be *stable* in case, for each $\varepsilon > 0$, there exists $\delta > 0$ such that, for each $x_1 \in \mathbb{R}^n$ with $\|x_1 - x_0\| < \delta$, the solution $x(t; t_0, x_1)$ exists on $[t_0, \infty)$ and satisfies $\|x(t; t_0, x_1) - x(t; t_0, x_0)\| < \varepsilon$ on $[t_0, \infty)$; i.e., a strong continuity property wrt x_0 is satisfied with t fixed at t_0.

Example 2.2. (1) Consider the scalar equation $x' = x^2$ on $[0, \infty) \times \mathbb{R}$. If $f(t, x) = x^2$, then $\frac{\partial f}{\partial x} = 2x$ and hence solutions of IVP's are unique. Consider the solution $x(t; 0, 0)$ of the IVP

$$\begin{cases} x' = x^2, \\ x(0) = 0. \end{cases}$$

By uniqueness of solutions of IVP's, $x(t; 0, 0) \equiv 0$. We claim that $x(t; 0, 0)$ is *not stable*.

To see this, let $\delta > 0$ be given and consider the solution $x(t; 0, \delta)$ of $x' = x^2$, $x(0) = \delta$. By separation of variables, $x(t; 0, \delta) = \frac{\delta}{1 - \delta t}$.

We see that, for any $\delta > 0$, $x(t; 0, \delta)$ doesn't exist on all of $[0, \infty)$ and in fact $|x(t; 0, 0) - x(t; 0, \delta)| \to +\infty$ as $t \to \frac{1}{\delta}^-$. See Figure 2.12.

However, we note that there is continuous dependence on initial conditions, provided we restrict ourselves to compact subintervals. Just select δ sufficiently small that $[0, \frac{1}{\delta}] \supset [a, b]$ and the continuity remarks above will be satisfied.

(2) Consider now the scalar equation $x' = -x$. The solution $x(t; 0, 0) \equiv 0$ is stable. To see this, solve the IVP: $x' = -x$, $x(0) = \delta$ and obtain $x(t; 0, \delta) = \delta e^{-t}$ which exists on $[0, \infty)$ and $|x(t; 0, \delta) - x(t; 0, 0)| \le$

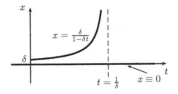

Fig. 2.12 $|x(t; 0, 0) - x(t; 0, \delta)| \to +\infty$ as $t \to \frac{1}{\delta}^-$.

$|\delta e^{-t}| \le \delta$. Choosing $\delta < \varepsilon$, we see that the definition concerning stability of $x(t; 0, 0) \equiv 0$ is satisfied.

(3) Consider

$$\begin{cases} x'' - 2tx' + t^2 x = 0, \\ x(0) = c_0, \quad x'(0) = c_1. \end{cases}$$

The equation satisfies Lipschitz condition on slabs of the form $[a, b] \times \mathbb{R}^2$. Let's look at this equation on say $[0, 2]$. We can write the corresponding integral equations

$$\begin{cases} x(t) = c_0 + \displaystyle\int_0^t x'(s) \, ds, \\ x'(t) = c_1 + \displaystyle\int_0^t x''(s) \, ds = c_1 + \displaystyle\int_0^t [2sx'(s) - s^2 x(s)] \, ds. \end{cases}$$

Suppose $y(t)$ is a solution of the same differential equation with $|y(0) - x(0)| + |y'(0) - x'(0)| < \delta$. Correspondingly, there are similar integral equations in terms of y. Now, choose y such that $y(0) \ne c_0$, $y'(0) \ne c_1$. Subtract corresponding integral equation expressions and thus obtain

$$y(t) - x(t) = y(0) - x(0) + \int_0^t [y'(s) - x'(s)] \, ds,$$

$$y'(t) - x'(t) = y'(0) - x'(0) + \int_0^t [2s[y'(s) - x'(s)] - s^2[y(s) - x(s)]] \, ds,$$

Hence,

$$|y(t) - x(t)| \le |y(0) - x(0)| + \int_0^t |y'(s) - x'(s)| \, ds, \qquad (2.4)$$

$$|y'(t) - x'(t)| \le |y'(0) - x'(0)| + \int_0^t [2s|y'(s) - x'(s)|$$
$$+ s^2 |y(s) - x(s)|] \, ds. \qquad (2.5)$$

Set $h(t) = |y(t) - x(t)| + |y'(t) - x'(t)|$, and then by (2.4) and (2.5), we have

$$h(t) \leq \delta + \int_0^t \left[(2s+1)|y'(s) - x'(s)| + s^2|y(s) - x(s)|\right] ds$$
$$\leq \delta + \int_0^t \left[(s^2 + 2s + 1)|y'(s) - x'(s)| + (s^2 + 2s + 1)|y(s) - x(s)|\right] ds$$
$$= \delta + \int_0^t (s^2 + 2s + 1)h(s)\, ds.$$

A special form of the Gronwall inequality states

$$h(t) \leq \delta + \left|\int_{t_0}^t K(s)h(s)\, ds\right|, \quad t_0 \leq t \leq \hat{t}$$

implies $h(t) \leq \delta e^{\left|\int_{t_0}^t K(s)\, ds\right|}$, for all $t \in [t_0, \hat{t}]$.

From the above form of the Gronwall inequality, $h(t) \leq \delta e^{\int_0^t (s^2 + 2s + 1)\, ds}$, for all $t \in [0, 2]$. Hence,

$$h(t) \leq \delta\, e^{\int_0^2 (s+1)^2\, ds} = \delta e^{\frac{26}{3}}.$$

Now, by our Lipschitz condition, the solution of the original IVP is unique on $[0, 2]$, and we see that the continuity wrt initial conditions in Remark 2.5 made above are satisfied again; at least as far as $y(t)$ and $[0, 2]$ are concerned, this is true, by taking $\delta < \varepsilon e^{-\frac{26}{3}}$. A simple modification can be made for any compact interval $[0, a]$.

Chapter 3

Smooth Dependence on Initial Conditions and Smooth Dependence on Parameters

3.1 Differentiation of Solutions wrt Initial Conditions

In the preceding chapter, we discussed the continuous dependence of solutions of IVP's wrt initial conditions and parameters. We now consider conditions under which solutions of IVP's can be differentiated wrt initial conditions and parameters, and we also examine properties of the resulting functions.

Theorem 3.1 (Differentiation wrt Initial Conditions). *Let $f(t,x)$ be continuous and have continuous first partial derivatives wrt the components of x on an open $D \subseteq \mathbb{R} \times \mathbb{R}^n$. Let $x(t;t_0,x_0) = x(t)$ be the unique solution of the IVP*

$$\begin{cases} x' = f(t,x), \\ x(t_0) = x_0 \end{cases}$$

and have maximal interval of existence (α, ω). Choose $[t_1, t_2] \subset (\alpha, \omega)$ and $t_0 \in [t_1, t_2]$. Let $A(t)$ be the Jacobian matrix, $A(t) = f_x(t, x(t))$ (i.e., $= f_x(t, x(t;t_0,x_0)))$. Then $x(t;t_0,x_0)$ has continuous partial derivatives wrt the components x_{0_j} of x_0, for $1 \leq j \leq n$, and wrt t_0 on $[t_1, t_2]$, hence on (α, ω). Furthermore, $\frac{\partial x(t;t_0,x_0)}{\partial x_{0_j}}$ is the solution of the IVP

$$\begin{cases} z' = A(t)z, \\ z(t_0) = e_j, \end{cases}$$

where $e_j = (\delta_{1j}, \delta_{2j}, \ldots, \delta_{nj})^T$ with

$$\delta_{ij} = \begin{cases} 1, & \text{if } i = j, \\ 0, & \text{if } i \neq j, \end{cases}$$

and $\frac{\partial x(t;t_0,x_0)}{\partial t_0}$ *is the solution of the IVP*

$$\begin{cases} z' = A(t)z, \\ z(t_0) = -f(t_0, x_0). \end{cases}$$

Proof. We consider first $\frac{\partial x(t;t_0,x_0)}{\partial x_{0_j}}$. Let $1 \leq j \leq n$ be fixed and let $x_m(t) = x(t; t_0, x_0 + \delta_m e_j)$, where $\{\delta_m\} \to 0$ (i.e., $x_m(t_0) = x_0 + \delta_m e_j$). Let $[t_1, t_2]$ be a fixed compact subinterval of (α, ω). Let $\eta > 0$ be such that $K \equiv \{(s, z) \mid t_1 \leq t \leq t_2, |s - t| \leq \eta$ and $\|z - x(t; t_0, x_0)\| \leq \eta\} \subseteq D$.

By the Kamke Theorem, there exists N_1 such that for all $m \geq N_1$, $[t_1, t_2] \subset (\alpha_m, \omega_m)$, the maximal interval of existence of $x_m(t)$, and the graph $\{(t, x_m(t)) \mid t_1 \leq t \leq t_2\} \subseteq K$, i.e., $\|x_m(t) - x(t)\| \leq \eta$, for $t_1 \leq t \leq t_2$, and $\lim_{m \to \infty} x_m(t) = x(t)$ uniformly on $[t_1, t_2]$.

Recall here the form of the Mean Value Theorem for vector valued functions, which we have seen earlier in Chapter 1: $f(t, y) - f(t, x) = \int_0^1 f_x(t, (1 - s)x + sy)(y - x)\, ds$, where f_x denotes the Jacobian matrix of f.

So, for all $m \geq N_1$,

$$x'_m(t) - x'(t) = f(t, x_m(t)) - f(t, x(t))$$

$$= \int_0^1 f_x(t,\, sx_m(t) + (1 - s)x(t))(x_m(t) - x(t))\, ds$$

$$= \int_0^1 f_x(t,\, sx_m(t) + (1 - s)x(t))\, ds\, (x_m(t) - x(t)). \quad (3.1)$$

Since $x_m(t) \to x(t)$ uniformly on $[t_1, t_2]$ and from the uniform continuity of f_x on K, it follows that as $m \to \infty$,

$$\int_0^1 f_x(t,\, sx_m(t) + (1 - s)x(t))\, ds \longrightarrow f_x(t,\, x(t)) \text{ uniformly on } [t_1, t_2].$$

Now, let

$$z_m(t) \equiv \frac{x_m(t) - x(t)}{\delta_m} = \frac{x_m(t; t_0, x_0 + \delta_m e_j) - x(t)}{\delta_m}.$$

Then, by (3.1),

$$z'_m(t) = \frac{x'_m(t) - x'(t)}{\delta_m} = \int_0^1 f_x(t, sx_m(t) + (1 - s)x(t))\, ds \cdot z_m.$$

Now, by construction,

$$z_m(t_0) = \frac{x_m(t_0) - x_0}{\delta_m} = \frac{x_0 + \delta_m e_j - x_0}{\delta_m} = e_j.$$

Moreover, $\left[\int_0^1 f_x(t, sx_m(t) + (1-s)x(t))\,ds\right]y$ is continuous on $[t_1, t_2] \times \mathbb{R}^n$, for all $m \geq N_1$, and converges uniformly to $f_x(t, x(t))y$ on each compact subset of $[t_1, t_2] \times \mathbb{R}^n$. By the Kamke Theorem, it follows that $\lim_{m \to \infty} z_m(t) = z(t)$ uniformly on the compact subinterval $[t_1, t_2]$, where $z(t)$ is the solution of the IVP

$$\begin{cases} z' = f_x(t, x(t))z, \\ z(t_0) = e_j. \end{cases}$$

Now let's look at $z(t)$ in detail. We see that $z_m(t) = \frac{x(t;t_0,x_0+\delta_m e_j)-x(t;t_0,x_0)}{\delta_m}$ is the type of difference quotient we would consider in looking for a partial derivative wrt the x_{0j} component. Above, we showed $\lim_{m \to \infty} z_m(t)$ existed (or $\lim_{\delta_m \to 0}$ of the difference quotient existed). Since this applies to any sequence $\{\delta_m\}$ with $\delta_m \to 0$ as $m \to \infty$, we can make the stronger statement that given $[t_1, t_2] \subset (\alpha, \omega)$ with $t_0 \in [t_1, t_2]$ and for each $\varepsilon > 0$, there exists $\delta > 0$ such that, for all $h \in \mathbb{R}$ with $|h| < \delta$, $[t_1, t_2]$ is contained in the maximal interval of existence of $x(t; t_0, x_0 + he_j)$ and

$$\left\| \frac{x(t;t_0,x_0+he_j) - x(t;t_0,x_0)}{h} - z(t) \right\| < \varepsilon,$$

that is, $z(t) \equiv \frac{\partial x(t;t_0,x_0)}{\partial x_{0j}}$. But $1 \leq j \leq n$ was arbitrary, thus partial derivatives wrt the components of the initial vector x_0 exist.

For differentiability wrt t_0, let $\delta_m \to 0$ and let $x_m(t)$ denote the solution $x(t; t_0 + \delta_m, x_0)$. As before, we have

$$x_m'(t) - x'(t) = \left[\int_0^1 f_x(t, sx_m(t) + (1-s)x(t))\,ds \right] \cdot [x_m(t) - x(t)].$$

Now define $z_m(t) \equiv \frac{x_m(t)-x(t)}{\delta_m}$. In this case,

$$z_m(t_0) = \frac{x(t_0; t_0 + \delta_m, x_0) - x(t_0; t_0, x_0)}{\delta_m}.$$

Note that $x(t_0) = x_0$ and $x_m(t_0 + \delta_m) = x_0$. In integral form,

$$x(t; t_0 + \delta_m, x_0) = x_0 + \int_{t_0+\delta_m}^t f(s, x(s; t_0 + \delta_m, x_0))\,ds,$$

Hence,

$$x(t_0; t_0 + \delta_m, x_0) = x_0 + \int_{t_0+\delta_m}^{t_0} f(s, x(s; t_0 + \delta_m, x_0))\,ds$$

$$= x_0 - \int_{t_0}^{t_0+\delta_m} f(s, x(s; t_0 + \delta_m, x_0))\,ds.$$

Consequently,

$$z_m(t_0) = -\frac{1}{\delta_m} \int_{t_0}^{t_0+\delta_m} f(s, x(s; t_0 + \delta_m, x_0))\,ds,$$

and as above, $z_m(t_0) \to -f(t_0, x_0)$. Then proceeding as in the first part of the proof with uniform convergence, etc., and with the sequence of initial conditions: $(t_0, z_m(t_0)) \to (t_0, -f(t_0, x_0))$, we obtain $\frac{\partial x(t; t_0, x_0)}{\partial t_0}$ is the solution of the IVP

$$\begin{cases} z' = f_x(t, x(t))z, \\ z(t_0) = -f(t_0, x_0). \end{cases} \qquad\qquad \square$$

Example 3.1. Consider the autonomous (independent of t) system

$$\begin{cases} f_1 = x_1' = x_1^2 + x_2, \\ f_2 = x_2' = x_1 + x_2^2, \\ x_1(t_0) = 0 = x_{0_1}, \\ x_2(t_0) = 0 = x_{0_2}. \end{cases}$$

Let $D = \mathbb{R} \times \mathbb{R}^2$. By the uniqueness of solutions of IVP's, solutions x_1, x_2 are continuous and differentiable functions of IC's. Here $x_1(t) \equiv 0$ and $x_2(t) \equiv 0$, i.e., $x(t) \equiv \begin{bmatrix} 0 \\ 0 \end{bmatrix}$. First compute

$$f_x\left(t, x(t)\right) = \begin{bmatrix} \frac{\partial f_1}{\partial x_1} & \frac{\partial f_1}{\partial x_2} \\ \frac{\partial f_2}{\partial x_1} & \frac{\partial f_2}{\partial x_2} \end{bmatrix}\Bigg|_{x(t)=\begin{bmatrix} 0 \\ 0 \end{bmatrix}} = \begin{bmatrix} 2x_1 & 1 \\ 1 & 2x_2 \end{bmatrix}\Bigg|_{\begin{cases} x_1 = 0 \\ x_2 = 0 \end{cases}} = \begin{bmatrix} 0 & 1 \\ 1 & 0 \end{bmatrix}.$$

By Theorem 3.1, $\dfrac{\partial x\left(t; t_0, \begin{bmatrix} 0 \\ 0 \end{bmatrix}\right)}{\partial x_{0_1}}$ is the solution to the IVP

$$z' = \begin{bmatrix} 0 & 1 \\ 1 & 0 \end{bmatrix} z, \quad z(t_0) = \begin{bmatrix} 1 \\ 0 \end{bmatrix} = e_1,$$

i.e.,

$$\begin{cases} z_1' = z_2, \\ z_2' = z_1, \end{cases} \qquad \begin{pmatrix} z_1 \\ z_2 \end{pmatrix}(t_0) = \begin{bmatrix} 1 \\ 0 \end{bmatrix}.$$

Hence,

$$\begin{cases} z_1'' = z_2' = z_1, \\ z_1(t_0) = 1, \quad z_1'(t_0) = z_2(t_0) = 0. \end{cases}$$

It follows that $z_1(t) = \cosh(t - t_0)$, $z_2(t) = \sinh(t - t_0)$. Hence,

$$\frac{\partial x\left(t; t_0, \begin{bmatrix} 0 \\ 0 \end{bmatrix}\right)}{\partial x_{0_1}} = \begin{bmatrix} \cosh(t - t_0) \\ \sinh(t - t_0) \end{bmatrix}.$$

Now we know that

$$\left\| \frac{x\left(t; t_0, he_1\right) - x\left(t; t_0, \left[\begin{smallmatrix}0\\0\end{smallmatrix}\right]\right)}{h} - \frac{\partial x\left(t; t_0, \left[\begin{smallmatrix}0\\0\end{smallmatrix}\right]\right)}{\partial x_{0_1}} \right\| \to 0, \text{ as } h \to 0.$$

Hence,

$$x\left(t; t_0, he_1\right) \approx h \begin{bmatrix} \cosh\left(t - t_0\right) \\ \sinh\left(t - t_0\right) \end{bmatrix},$$

where the initial condition is $x_1(t_0) = h$, $x_2(t_0) = 0$.

3.2 Differentiation of Solutions wrt Parameters

Suppose now that $f(t, x, \lambda)$ is continuous and has continuous first partials wrt the components of x and λ on an open set $D \subseteq \mathbb{R} \times \mathbb{R}^n \times \mathbb{R}^m$. Consider the IVP

$$\begin{cases} x' = f(t, x, \lambda_0), \\ x(t_0) = x_0 \end{cases}$$

and let $x(t; t_0, x_0, \lambda_0)$ denote the solution. We wish to discuss $\frac{\partial x(t; t_0, x_0, \lambda_0)}{\partial \lambda_{0_j}}$, for $1 \leq j \leq m$.

Change the IVP to the nonparametric situation by $z = \left[\begin{smallmatrix}x\\\lambda\end{smallmatrix}\right]$, so that

$$z' = \begin{bmatrix} f(t, z) \\ 0 \end{bmatrix} = h(t, z), \quad z(t_0) = \begin{bmatrix} x_0 \\ \lambda_0 \end{bmatrix}.$$

By Theorem 3.1, $\frac{\partial z(t; t_0, z_0)}{\partial \lambda_{0_j}}$ is the solution of the IVP $y' = h_z(t, z(t; t_0, z_0)) \cdot y$ satisfying

$$y(t_0) = \left.\begin{bmatrix} \left.\begin{matrix} 0 \\ \vdots \\ 0 \end{matrix}\right\} n \text{ components} \\ 0 \\ \left.\begin{matrix} \vdots \\ 0 \\ 1 \\ 0 \\ \vdots \\ 0 \end{matrix}\right\} m \text{ components} \end{bmatrix}\right. = e_{n+j}, \quad \text{where} \quad h(t, z) = \begin{bmatrix} f_1 \\ \vdots \\ f_n \\ 0 \\ \vdots \\ 0 \end{bmatrix}$$

and the Jacobian

$$h_z(t, z(t; t_0, z_0))$$

$$= \left[\begin{array}{ccccccc}
\frac{\partial f_1}{\partial x_1} & \cdots & \frac{\partial f_1}{\partial x_n} & \frac{\partial f_1}{\partial \lambda_1} & \cdots & \frac{\partial f_1}{\partial \lambda_m} \\
\vdots & \ddots & \vdots & \vdots & \ddots & \vdots \\
\frac{\partial f_n}{\partial x_1} & \cdots & \frac{\partial f_n}{\partial x_n} & \frac{\partial f_n}{\partial \lambda_1} & \cdots & \frac{\partial f_n}{\partial \lambda_m} \\
0 & \cdots & 0 & 0 & \cdots & 0 \\
\vdots & \ddots & \vdots & \vdots & \ddots & \vdots \\
0 & \cdots & 0 & 0 & \cdots & 0
\end{array}\right]_{z(t; t_0, z_0)}$$

$$= \left[\begin{array}{c|c}
\left[f_x(t, x(t; t_0, x_0, \lambda_0), \lambda_0)\right]_{n \times n} & \left[f_\lambda(t, x(t; t_0, x_0, \lambda_0), \lambda_0)\right]_{n \times m} \\
\hline
0_{m \times n} & 0_{m \times m}
\end{array}\right].$$

If $\frac{\partial z(t; t_0, z_0)}{\partial \lambda_{0_j}}$ is the solution of IVP

$$\begin{cases} y' = h_z(t, z(t; t_0, z_0))y, \\ y(t_0) = e_{n+j}, \end{cases}$$

then the last m components satisfy the IVP's,

$$\begin{cases} y'_{n+k} = 0, & 1 \le k \le m, \\ y_{n+k}(t_0) = 0, \ k \ne j, & y_{n+j}(t_0) = 1, \end{cases}$$

whose solutions are $y_{n+k}(t) \equiv 0$, for $1 \le k \le m$, $k \ne j$, and $y_{n+j}(t) \equiv 1$. Thus, we have that $\frac{\partial x(t; t_0, x_0, \lambda_0)}{\partial \lambda_{0_j}}$ is the solution of the IVP

$$\begin{cases} u' = f_x(t, x(t; t_0, x_0, \lambda_0), \lambda_0)u + h_j(t), \\ u(t_0) = 0, \end{cases} \tag{3.2}$$

where

$$h_j(t) = \left[\begin{array}{c}
\frac{\partial f_1}{\partial \lambda_j}(t, x(t; t_0, x_0, \lambda_0), \lambda_0) \\
\frac{\partial f_2}{\partial \lambda_j}(t, x(t; t_0, x_0, \lambda_0), \lambda_0) \\
\vdots \\
\frac{\partial f_n}{\partial \lambda_j}(t, x(t; t_0, x_0, \lambda_0), \lambda_0)
\end{array}\right].$$

Theorem 3.2 (Differentiation wrt Parameters). *Suppose $f(t, x, \lambda)$ is continuous and has continuous first partial derivatives wrt the components* *of x and λ on an open set $D \subseteq \mathbb{R} \times \mathbb{R}^n \times \mathbb{R}^m$. Let $(t_0, x_0, \lambda_0) \in D$ and let $x(t; t_0, x_0, \lambda_0)$ be the unique solution of*

$$\begin{cases} x' = f(t, x, \lambda_0), \\ x(t_0) = x_0. \end{cases}$$

Then $x(t; t_0, x_0, \lambda_0)$ has partial derivatives wrt the components of λ_0 on $(\alpha(t_0, x_0, \lambda_0), \omega(t_0, x_0, \lambda_0))$ and $\frac{\partial x(t; t_0, x_0, \lambda_0)}{\partial \lambda_{0j}}$ is the solution of (3.2).

Remark 3.1. Consider the case of IVP's for a linear system

$$\begin{cases} x' = A(t)_{n \times n}\, x + h(t)_{n \times 1} = f(t, x), \\ x(t_0) = x_0, \end{cases}$$

where

$$A(t) = \begin{bmatrix} a_{11}(t) & \cdots & a_{1n}(t) \\ \vdots & \ddots & \vdots \\ a_{n1}(t) & \cdots & a_{nn}(t) \end{bmatrix},$$

$$f(t, x) = \begin{bmatrix} a_{11}(t)\, x_1 + \cdots + a_{1n}(t) x_n + h_1(t) \\ \vdots \\ a_{n1}(t)\, x_1 + \cdots + a_{nn}(t) x_n + h_n(t) \end{bmatrix} = \begin{bmatrix} f_1(t, x) \\ \vdots \\ f_n(t, x) \end{bmatrix}.$$

Hence, the Jacobian is

$$f_x(t, x(t; t_0, x_0))$$

$$= \begin{bmatrix} \frac{\partial f_1}{\partial x_1} & \cdots & \frac{\partial f_1}{\partial x_n} \\ \vdots & \ddots & \vdots \\ \frac{\partial f_n}{\partial x_1} & \cdots & \frac{\partial f_n}{\partial x_n} \end{bmatrix} \Bigg|_{x(t; t_0, x_0)} = \begin{bmatrix} a_{11}(t) & \cdots & a_n(t) \\ \vdots & \ddots & \vdots \\ a_{n1}(t) & \cdots & a_{nn}(t) \end{bmatrix} = A(t),$$

i.e., $f_x(t, x(t; t_0, x_0)) = A(t)$, for all solutions $x(t)$ of the original system of differential equations.

Definition 3.1. Given a solution $x(t; t_0, x_0)$ of $x' = f(t, x)$, $x(t_0) = x_0$, the differential equation $y' = f_x(t, x(t; t_0, x_0))y$ is called the *first variational equation* of $x' = f(t, x)$ wrt the solution $x(t; t_0, x_0)$.

Remark 3.2. Consider the IVP for the nth order scalar equation,

$$\begin{cases} x^{(n)} = f(t, x, x', \ldots, x^{(n-1)}), \\ x^{(i-1)}(t_0) = c_i, \quad 1 \le i \le n, \end{cases}$$

where $f(t, y_1, y_2, \ldots y_n) : I \times \mathbb{R}^n \to \mathbb{R}$ is continuous.

We transform this into the first order system

$$\begin{cases} y' = \tilde{f}(t, y), \\ y(t_0) = c = \begin{bmatrix} c_1 \\ \vdots \\ c_n \end{bmatrix}, \end{cases}$$

that is,

$$\begin{bmatrix} y_1 \\ \vdots \\ y_n \end{bmatrix}' = \begin{bmatrix} y_2 \\ \vdots \\ y_n \\ f(t, y_1, \ldots, y_n) \end{bmatrix} = \begin{bmatrix} f_1(t, y) \\ \vdots \\ f_n(t, y) \end{bmatrix},$$

where

$$y_1 = x,$$
$$y_1' = y_2 = x',$$
$$\vdots$$
$$y_{n-1}' = y_n = x^{(n-1)},$$
$$y_n' = f(t, y_1, \ldots, y_n).$$

For this first order system, we have

$$\tilde{f}_y(t, y(t)) = \begin{bmatrix} 0 & 1 & 0 & 0 & \cdots & 0 & 0 \\ 0 & 0 & 1 & 0 & \cdots & 0 & 0 \\ 0 & 0 & 0 & 1 & \cdots & 0 & 0 \\ \vdots & \vdots & \vdots & \vdots & \ddots & \vdots & \vdots \\ 0 & 0 & 0 & 0 & \cdots & 1 & 0 \\ 0 & 0 & 0 & 0 & \cdots & 0 & 1 \\ \frac{\partial f}{\partial y_1} & \frac{\partial f}{\partial y_2} & \frac{\partial f}{\partial y_3} & \frac{\partial f}{\partial y_4} & \cdots & \frac{\partial f}{\partial y_{n-1}} & \frac{\partial f}{\partial y_n} \end{bmatrix}\Bigg|_{y(t)}.$$

It follows from Theorem 3.1 that $\frac{\partial y(t;t_0,c)}{\partial c_j}$ is the solution of the IVP

$$\begin{cases} z' = \begin{bmatrix} 0 & 1 & 0 & \cdots & 0 \\ 0 & 0 & 1 & \cdots & 0 \\ \vdots & \vdots & \vdots & \ddots & \vdots \\ 0 & 0 & 0 & \cdots & 1 \\ \frac{\partial f}{\partial y_1} & \frac{\partial f}{\partial y_2} & \frac{\partial f}{\partial y_3} & \cdots & \frac{\partial f}{\partial y_n} \end{bmatrix}\Bigg|_{(t,y(t;t_0,c))} \cdot z, \\ z(t_0) = e_j = (\delta_{1j}, \delta_{2j}, \ldots, \delta_{nj})^T, \end{cases}$$

which is, in component-wise form,

$$\begin{cases} z_1' = z_2, \\ z_2' = z_3, \\ \quad \vdots \\ z_{n-1}' = z_n, \\ z_n' = \frac{\partial f}{\partial y_1}\left(t, y\left(t; t_0, c\right)\right) z_1 + \frac{\partial f}{\partial y_2}\left(t, y\left(t; t_0, c\right)\right) z_2 \\ \qquad + \cdots + \frac{\partial f}{\partial y_n}\left(t; y\left(t; t_0, c\right)\right) z_n. \end{cases}$$

Equivalently, we conclude that $z_1(t)$, the first component of the solution of the IVP for this system, is the solution of the nth order IVP (with $z_1 = z$)

$$\begin{cases} z^{(n)}(t) = \frac{\partial f}{\partial y_1}(t, x(t; t_0, c), \ldots, x^{(n-1)}(t; t_0, c))z \\ \qquad + \frac{\partial f}{\partial y_2}(t, x(t; t_0, c), \ldots, x^{(n-1)}(t; t_0, c))z' \\ \qquad + \cdots \\ \qquad + \frac{\partial f}{\partial y_n}(t, x(t; t_0, c), \ldots, x^{(n-1)}(t; t_0, c))z^{(n-1)}, \\ z^{(i-1)}(t_0) = \delta_{ij}, \quad 1 \le i \le n. \end{cases} \qquad (3.3)$$

But $z_1(t)$ is the first component of $\frac{\partial y(t; t_0, c)}{\partial c_j}$. So, $z_1(t) = \frac{\partial y_1(t; t_0, c)}{\partial c_j}$, and in particular $z_1(t) = \frac{\partial x(t; t_0, c)}{\partial c_j} = \frac{\partial x(t; t_0, c_1, \ldots, c_n)}{\partial c_j}$ is the solution of the nth order linear differential equation in (3.3) and satisfies the IC's there.

Note: In (3.3), the coefficient of $z^{(j)}$ is just a continuous function of t.

Example 3.2. This example illustrates Remark 3.2. Let $f(t, y_1, y_2) = y_1^3 - y_1$ and consider

$$\begin{cases} x'' = x^3 - x, \\ x(0) = 1, \ x'(0) = 0, \quad \implies \quad c = \begin{bmatrix} 1 \\ 0 \end{bmatrix}. \end{cases}$$

Now $\frac{\partial f}{\partial y_1}, \frac{\partial f}{\partial y_2}$ are continuous and hence the IVP has a unique solution. By inspection, our unique solution of the given nonlinear IVP is $x(t; 0, c) = x(t; 0, 1, 0) \equiv 1$. We wish to calculate $\frac{\partial x(t; 0, c_1, c_2)}{\partial c_1}\Big|_{\{c_1 = 1 \atop c_2 = 0}}$.

From above, $\frac{\partial x(t;0,c_1,c_2)}{\partial c_1}\Big|_{\{^{c_1=1}_{c_2=0}}}$ is the solution of the IVP

$$
\begin{cases}
z'' = \dfrac{\partial f}{\partial y_1}\left(t, x(t;0,1,0), x'(t;0,1,0)\right) z \\[2mm]
\qquad + \dfrac{\partial f}{\partial y_2}\left(t, x(t;0,1,0), x'(t;0,1,0)\right)z, \\[2mm]
z(0) = 1, \quad z'(0) = 0,
\end{cases}
$$

i.e., $\frac{\partial x(t;0,c_1,c_2)}{\partial c_1}\Big|_{\{^{c_1=1}_{c_2=0}}}$ is the solution of

$$
\begin{cases}
z'' = \left[3(x(t;0,1,0))^2 - 1\right]z + 0 = (3(1)^2 - 1)z = 2z, \\[2mm]
z(0) = 1, \quad z'(0) = 0.
\end{cases}
$$

Hence

$$
\frac{\partial x\left(t; 0, c_1, c_2\right)}{\partial c_1}\Bigg|_{\{^{c_1=1}_{c_2=0}}} = \cosh\sqrt{2}t.
$$

Now calculate $\frac{\partial x(t;0,c_1,c_2)}{\partial c_2}\Big|_{\{^{c_1=1}_{c_2=0}}}$. In this case, we seek the solution of

$$
\begin{cases}
z'' = [3(x(t;0,1,0)^2 - 1]z, \\[2mm]
z(0) = 0, \quad z'(0) = 1.
\end{cases}
$$

So, $\frac{\partial x(t;0,c_1,c_2)}{\partial c_1}\Big|_{\{^{c_1=1}_{c_2=0}}} = \frac{1}{\sqrt{2}}\sinh\sqrt{2}t.$

Note: Above we use $\frac{\partial x(t;0,c_1,c_2)}{\partial c_1}\Big|_{\{^{c_1=1}_{c_2=0}}}$ rather than $\frac{\partial x(t;0,1,0)}{\partial c_1}\Big|_{\{^{c_1=1}_{c_2=0}}}$ to avoid confusion. We could have just recalled that

$$
\frac{\partial x\left(t; 0, c_1, c_2\right)}{\partial c_1}\Bigg|_{\{^{c_1=1}_{c_2=0}}} = \lim_{h\to 0}\frac{x\left(t; 0, 1+h, 0\right) - x\left(t; 0, 1, 0\right)}{h}.
$$

$\boxed{\text{Exercise}}$ **17.** **A.** Let λ be an m-vector and let $x(t; t_0, c, \lambda)$ be the solution of the IVP

$$
\begin{cases}
x^{(n)} = f\left(t, x, x', \ldots, x^{(n-1)}, \lambda\right), \\[2mm]
x^{(i-1)}(t_0) = c_i, \quad 1 \le i \le n.
\end{cases}
$$

(So $c = (c_1, c_2, \ldots, c_n)^T$.) Determine the IVP which has $\frac{\partial x(t;\, t_0,\, c,\, \lambda_0)}{\partial \lambda_j}$ as a solution.

B. Apply problem A to each of the following:
   (a) $x'' + \lambda x = 0$, $x(0) = 0$, $x'(0) = 1$, $\lambda_0 = 4$.
   (b) $(1-t^2)x'' - 2tx' + \lambda(\lambda+1)x = 0$, $t_0 = 0$, $\lambda = n$; solution is $P_n(t)$.
   (c) $t^2 x'' + tx' + (t^2 - \lambda^2)x = 0$, $t_0 = 1$, $\lambda = 0$; solution is $J_0(t)$.
C. In B(a), calculate $\frac{\partial x}{\partial \lambda}$ by solving the IVP, then differentiating wrt λ.
D. Given the scalar IVP

$$\begin{cases} x' = 1 + x^2, \\ x(0) = \eta, \end{cases}$$

compute $\frac{\partial x(t;0,\eta)}{\partial \eta}$ and $\frac{\partial x(t;0,\eta)}{\partial t_0}$, by applying Theorem 3.1, and also by solving the IVP, then differentiating the solution.
E. Given the scalar equation $x' = \lambda + \cos x$, $\lambda \in \mathbb{R}$, compute $\frac{\partial x(t;0,0,0)}{\partial \lambda}$ by applying Theorem 3.2.

3.3   Maximal Solutions and Minimal Solutions

For a final application of the Kamke Theorem, we have the following.

Definition 3.2. Let $\varphi(t,x)$ be continuous on an open set $D \subseteq \mathbb{R} \times \mathbb{R}$ and let $(t_0, x_0) \in D$. A solution $x(t)$ of

$$\begin{cases} x' = \varphi(t,x), \\ x(t_0) = x_0, \end{cases}$$

on a maximal interval of existence (α_0, ω_0) is said to be a *maximal solution* in case, for any other solution $y(t)$ on a maximal interval (α_1, ω_1), we have $x(t) \geq y(t)$ on $(\alpha_0, \omega_0) \cap (\alpha_1, \omega_1)$. A *minimal solution* is similarly defined.

Note: $\varphi(t,x)$ is real-valued; i.e., $\varphi(t,x) : D \to \mathbb{R}$.

Exercise 18. Show that a maximal solution of

$$\begin{cases} x' = \varphi(t,x), \\ x(t_0) = x_0, \end{cases}$$

is unique.

Theorem 3.3. *Let $\varphi(t,x)$ be continuous on an open set $D \subseteq \mathbb{R} \times \mathbb{R}$ and let $(t_0, x_0) \in D$. Then the IVP*

$$\begin{cases} x' = \varphi(t,x), \\ x(t_0) = x_0, \end{cases} \tag{3.4}$$

has a maximal solution $x_M(t)$ and a minimal solution $x_m(t)$.

Proof. For each $n \geq 1$, let $u_n(t)$ denote a solution of the IVP

$$\begin{cases} x' = \varphi(t,x) + \dfrac{1}{n}, \\ x(t_0) = x_0, \end{cases}$$

defined on a maximal interval of existence (α_n, ω_n). Then by the Kamke Theorem, there is a solution $x^*(t)$ of (3.4) on a maximal interval of existence (α^*, ω^*) and a subsequence $\{u_{n_k}(t)\}$ which converges uniformly to $x^*(t)$ on each compact subinterval of (α^*, ω^*).

Assume now that $y(t)$ is another solution of (3.4) on a maximal interval $(\bar{\alpha}, \bar{\omega})$ such that, for some $t_1 \in (\alpha^*, \omega^*) \cap (\bar{\alpha}, \bar{\omega})$ and $t_1 > t_0$, $y(t_1) > x^*(t_1)$. Now $u_{n_k}(t) \to x^*(t)$ on $[t_0, t_1]$ as $k \to \infty$ and so there exists $\{u_{n_{k_0}}\}$ such that $u_{n_{k_0}}(t_1) < y(t_1)$. Since $u_{n_{k_0}}(t_0) = x_0 = y(t_0)$, it follows that there is

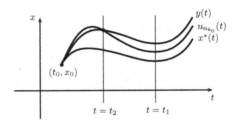

Fig. 3.1   The solutions $x^*(t)$, $y(t)$, and $u_{n_{k_0}}(t)$.

some t_2 with $t_0 \leq t_2 \leq t_1$ such that $y(t_2) = u_{n_{k_0}}(t_2)$ and $y(t) > u_{n_{k_0}}(t)$ on $(t_2, t_1]$. See Figure 3.1. Thus

$$\frac{d}{dt}\left[y(t) - u_{n_{k_0}}(t) \right]\Big|_{t=t_2} \geq 0.$$

Therefore,

$$u'_{n_{k_0}}(t_2) \leq y'(t_2) = \varphi(t_2, y(t_2)) = \varphi(t_2, u_{n_{k_0}}(t_2)) = u'_{n_{k_0}}(t_2) - \frac{1}{n_{k_0}},$$

which is a contradiction. Thus, no such $y(t)$ exists and it follows that $x^*(t)$ **is a maximal solution** on $[t_0, \omega^*)$. A similar argument shows that $x^*(t)$ **is a minimal solution** on $(\alpha^*, t_0]$.

Now, for each $n \geq 1$, let $\omega_n(t)$ be a solution of

$$\begin{cases} x' = \varphi(t,x) - \dfrac{1}{n}, \\ x(t_0) = x_0, \end{cases}$$

on a maximal interval (α^n, ω^n). Again, by the Kamke Theorem, there exists a solution $x^{**}(t)$ of (3.4), on a maximal interval $(\alpha^{**}, \omega^{**})$ and a subsequence $\{\omega_{n_j}(t)\}$ which converges uniformly to $x^{**}(t)$ on each compact subinterval of $(\alpha^{**}, \omega^{**})$. Arguing as above, it can be shown that x^{**} **is a minimal solution of** (3.4) **on** $[t_0, \omega^{**})$ and **a maximal solution on** $(\alpha^{**}, t_0]$. Hence

$$x_M(t) := \begin{cases} x^*(t), & t \in [t_0, \omega^*), \\ x^{**}(t), & t \in (\alpha^{**}, t_0], \end{cases}$$

with maximal interval (α^{**}, ω^*), and

$$x_m(t) := \begin{cases} x^{**}(t), & t \in [t_0, \omega^{**}), \\ x^*(t), & t \in (\alpha^*, t_0], \end{cases}$$

with maximal interval (α^*, ω^{**}).

(Note: x_M and x_m are unique.) □

Chapter 4

Some Comparison Theorems and Differential Inequalities

4.1 Comparison Theorems and Differential Inequalities

For a function $y(t)$, defined in some neighborhood of the point t_0, the *Dini derivatives* at the point t are defined as follows:

$$D^+y(t) = \varlimsup_{h \to 0^+} \frac{y(t+h) - y(t)}{h} \quad \text{(upper-right derivative)},$$

$$D_+y(t) = \varliminf_{h \to 0^+} \frac{y(t+h) - y(t)}{h} \quad \text{(lower-right derivative)},$$

$$D^-y(t) = \varlimsup_{h \to 0^-} \frac{y(t+h) - y(t)}{h} \quad \text{(upper-left derivative)},$$

$$D_-y(t) = \varliminf_{h \to 0^-} \frac{y(t+h) - y(t)}{h} \quad \text{(lower-left derivative)}.$$

For the next few results, we will be concerned with differential inequalities. In establishing these results, we will make use of maximal and minimal solutions.

Theorem 4.1. *Let $\varphi(t, x)$ be continuous on an open set $D \subseteq \mathbb{R} \times \mathbb{R}$ and $(t_0, x_0) \in D$. Let $x_M(t)$ be the maximal solution of*

$$\begin{cases} x' = \varphi(t, x), \\ x(t_0) = x_0, \end{cases}$$

on a maximal interval (α, ω). Let $v(t)$ be continuous and real-valued on $[t_0, t_0 + a]$ and satisfy the following:

(1) $(t, v(t)) \in D$, *for* $t \in [t_0, t_0 + a]$,

(2) $D^+v(t) \le \varphi(t, v(t))$, *for* $t \in [t_0, t_0 + a)$, *and*

(3) $v(t_0) \leq x_0$.

Then, $v(t) \leq x_M(t)$ on $[t_0, t_0 + a] \cap (\alpha, \omega)$.

Proof. Assume the conclusion to be false. Then, there exists $t_1 > t_0$ with $t_1 \in [t_0, t_0 + a] \cap (\alpha, \omega)$ such that $v(t_1) > x_M(t_1)$. Now $v(t_0) \leq x_0 = x_M(t_0)$ and so, as in Theorem 3.3, there exists $n \geq 1$ and a solution $u_n(t)$ of

$$\begin{cases} x' = \varphi(t, x) + \dfrac{1}{n}, \\[2mm] x(t_0) = x_0, \end{cases}$$

on $[t_0, t_1]$ such that $u_n(t_1) < v(t_1)$.

As in Theorem 3.3, there exists a t_2 with $t_0 \leq t_2 < t_1$ such that $u_n(t_2) = v(t_2)$ and $u_n(t) < v(t)$ on $(t_2, t_1]$. Then

$$\varliminf_{h \to 0^+} \frac{[v(t_2 + h) - v(t_2)] - [u_n(t_2 + h) - u_n(t_2)]}{h}$$

$$= \varliminf_{h \to 0^+} \frac{v(t_2 + h) - u_n(t_2 + h)}{h} \geq 0.$$

Hence, $D^+[v(t) - u_n(t)]|_{t=t_2} \geq 0$. Now, u_n is differentiable, consequently

$$0 \leq D^+[v(t) - u_n(t)]|_{t=t_2}$$
$$= D^+ v(t_2) - u_n'(t_2)$$
$$\leq \varphi(t_2, v(t_2)) - \left[\varphi(t_2, u_n(t_2)) + \frac{1}{n}\right]$$
$$= \varphi(t_2, v(t_2)) - \varphi(t_2, u_n(t_2)) - \frac{1}{n}$$
$$= -\frac{1}{n} < 0,$$

which is a contradiction. Therefore, our assumption is false, and it follows that $v(t) \leq x_M(t)$ on $[t_0, t_0 + a] \cap (\alpha, \omega)$. $\qquad\square$

Exercise **19.** In each of the following, modify the hypotheses of Theorem 4.1 concerning $v(t)$ in the indicated way, and prove (some of) the corresponding results.

(1) $D_+ v(t) \geq \varphi(t, v(t))$ on $[t_0, t_0 + a]$, $v(t_0) \geq x_0 \Rightarrow v(t) \geq x_m(t)$.

(2) $D^- v(t) \geq \varphi(t, v(t))$ on $(t_0 - a, t_0]$, $v(t_0) \leq x_0 \Rightarrow v(t) \leq x_M(t)$.

(3) $D_- v(t) \leq \varphi(t, v(t))$ on $(t_0 - a, t_0]$, $v(t_0) \geq x_0 \Rightarrow v(t) \geq x_m(t)$.

Corollary 4.1 (To Theorem 4.1**).** *If $v(t)$ is continuous on $[a, b]$ and $D^+ v(t) \leq 0$ on $[a, b]$, then $v(t)$ is nonincreasing on $[a, b]$.*

Proof. In Theorem 4.1, take $\varphi(t, x) \equiv 0$. Let $a \le t_0 < t_1 \le b$ and consider the IVP

$$\begin{cases} x' = \varphi(t, x) = 0, \\ x(t_0) = v(t_0). \end{cases}$$

It follows that $x_M(t) = v(t_0)$, for all $t \in [a, b]$ and that $v(t) \le v(t_0)$, for all $t_0 \le t \le b$. In particular, $v(t_1) \le v(t_0)$. $\qquad\square$

Corollary 4.2. *If $\psi(t, x)$ and $\varphi(t, x)$ are continuous real-valued on an open set $D \subseteq \mathbb{R} \times \mathbb{R}$ such that $\psi(t, x) \le \varphi(t, x)$ on D, if $x_M(t)$ is the maximal solution of*

$$\begin{cases} x' = \varphi(t, x), \\ x(t_0) = x_0, \end{cases}$$

where $(t_0, x_0) \in D$, and if $v(t)$ is a solution of $x' = \psi(t, x)$ with $v(t_0) \le x_0$, then $v(t) \le x_M(t)$ on a common right interval of existence.

Proof. We have that $v'(t) = \psi(t, v(t)) \le \varphi(t, v(t))$ for all t in the maximal interval for $v(t)$. Moreover, $v(t_0) \le x_0$. It follows from Theorem 4.1 that $v(t) \le x_M(t)$ on a common right interval of existence. $\qquad\square$

$\boxed{\text{Exercise}}$ **20.** Using Corollary 4.2, prove that the solution of the IVP

$$\begin{cases} x' = t^2 + x^2, \\ x(0) = 0 \end{cases}$$

has a vertical asymptote somewhere between $\frac{\pi}{2}$ and $\sqrt{2\pi}$.

Hint: In conjunction with Corollary 4.2, use the comparison equations $x' = t^2 + 2tx + x^2$ and $x' = a^2 + x^2$, where $a > 0$.

The solution of the IVP resembles Figure 4.1.

If $v(t)$ is the solution, observe that $v'(0) = 0^2 + 0^2 = 0$, but $v'(t) = t^2 + v^2 > 0$, for $t > 0$ and hence $v(t) > 0$, for $t > 0$.

A possible way to use the hints is as follows: First let $v(t)$ be a solution of $x' = t^2 + x^2$. Then $v'(t) = t^2 + v^2(t) \le t^2 + 2tv(t) + v^2(t)$, for $t \ge 0$. Find the solution of $x' = t^2 + 2tx + x^2$, $x(0) = 0$, apply Corollary 4.2 and show that $v(t)$ has an asymptote greater than $\frac{\pi}{2}$.

For the second inequality consider $x' = a^2 + x^2$, $a > 0$. As above let $v(t)$ be the solution of $x' = t^2 + x^2$, $x(0) = 0$ (recall then $v(a) > 0$), and notice that $v'(t) = t^2 + v^2(t) \ge a^2 + v^2$, for $t \ge a$. So find the solution of

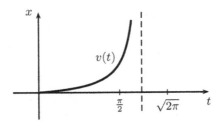

Fig. 4.1 The solution of the IVP resembles $v(t)$.

$x' = a^2 + x^2$, $x(a) =$ probably 0. If a is picked correctly, an application of Corollary 4.2 will show that $v(t)$ has an asymptote less than $\sqrt{2\pi}$.

Remark 4.1. If $x(t) : [a,b] \to \mathbb{R}^n$ is differentiable on $[a,b)$, then $D^+\|x(t)\| \le \|x'(t)\|$ on $[a,b)$.

Proof. Let $t \in [a,b)$ and $h > 0$ such that $t+h \in [a,b)$. Then

$$\|x(t+h)\| - \|x(t)\| \le \|x(t+h) - x(t)\|,$$

which yields

$$\frac{\|x(t+h)\| - \|x(t)\|}{h} \le \frac{1}{h}\|x(t+h) - x(t)\| = \left\|\frac{x(t+h) - x(t)}{h}\right\|,$$

for $h > 0$. The above inequality is true for all $h > 0$ such that $t+h \in [a,b)$. Hence,

$$\lim_{h\downarrow 0}\frac{\|x(t+h)\| - \|x(t)\|}{h} \le \lim_{h\downarrow 0}\left\|\frac{x(t+h) - x(t)}{h}\right\| = \left\|\lim_{h\downarrow 0}\frac{x(t+h) - x(t)}{h}\right\|,$$

implying, by continuity,

$$D^+\|x(t)\| \le \|x'(t)\|. \qquad \square$$

For notational purposes, let D_r denote the right-hand derivative of a function. If D_r exists, then $D_r = D^+ = D_+$.

Lemma 4.1. *If $x(t)$ is a differentiable vector-valued function on $[a,b]$, then the right-hand derivative, $D_r\|x(t)\|$, exits on $[a,b)$ and satisfies $D_r\|x(t)\| \le \|x'(t)\|$.*

Proof. Let $x, u \in \mathbb{R}^n$ be fixed and let $0 < \theta \leq 1$ and $h > 0$. Then

$$\|x + \theta h u\| - \underbrace{\|\theta x + \theta h u\|}_{\theta \|x + h u\|} \leq \|x - \theta x\| = (1 - \theta)\|x\| = \|x\| - \theta \|x\|.$$

Hence

$$\frac{\|x + \theta h u\| - \|x\|}{\theta} \leq \|x + h u\| - \|x\|,$$

and so

$$\frac{\|x + \theta h u\| - \|x\|}{\theta h} \leq \frac{\|x + h u\| - \|x\|}{h}.$$

It follows that $\frac{\|x + h u\| - \|x\|}{h}$ decreases, as $h \to 0^+$; i.e., the quotient $\frac{\|x + k u\| - \|x\|}{k}$ is a nondecreasing function in $k \in (0, h]$. To see this, let $0 < k < h$, so that $k = \theta h$, where $\theta = \frac{k}{h}$. Then

$$\frac{\|x + k u\| - \|x\|}{k} = \frac{\|x + \theta h u\| - \|x\|}{\theta h} \leq \frac{\|x + h u\| - \|x\|}{h},$$

thus the claim is true. (Furthermore, for $h \to 0^-$, a reversal of the inequalities occurs and $\frac{\|x + h u\| - \|x\|}{h} \leq \frac{\|x + k u\| - \|x\|}{k}$ for $h < k < 0$.)

Now

$$\|\|x + h u\| - \|x\|\| \leq \|x + h u - x\| = h\|u\|, \quad \text{for } h > 0.$$

Then,

$$-h\|u\| \leq \|x + h u\| - \|x\| \leq h\|u\|, \quad \text{for } h > 0,$$

and so

$$-\|u\| \leq \frac{\|x + h u\| - \|x\|}{h} \leq \|u\|, \quad \text{for } h > 0.$$

Here, the nondecreasing function is bounded below by $-\|u\|$ (and above by $\|u\|$). So, $\lim_{h \downarrow 0} \frac{\|x + h u\| - \|x\|}{h}$ exists. In particular, if $t \in [a, b)$ is fixed then $\lim_{h \downarrow 0} \frac{\|x(t) + h x'(t)\| - \|x(t)\|}{h}$ exists.

Note here that

$$\left| (\|x(t + h)\| - \|x(t)\|) - (\|x(t) + h x'(t)\| - \|x(t)\|) \right|$$
$$= \left| \|x(t + h)\| - \|x(t) + h x'(t)\| \right|$$
$$\leq \|x(t + h) - x(t) - h x'(t)\|.$$

Thus, for $h > 0$ such that $t + h \in [a, b)$, we have

$$0 \leq \left| \frac{\|x(t + h)\| - \|x(t)\|}{h} - \frac{\|x(t) + h x'(t)\| - \|x(t)\|}{h} \right|$$
$$\leq \left\| \frac{x(t + h) - x(t)}{h} - x'(t) \right\|.$$

Now as $h \to 0$, the right-hand side approaches 0, thus the left-hand side approaches 0. So, if $h \downarrow 0$, since $\lim_{h\downarrow 0} \frac{\|x(t)+hx'(t)\| - \|x(t)\|}{h}$ exists from above, we have $\lim_{h\downarrow 0} \frac{\|x(t+h)\| - \|x(t)\|}{h}$ exists and

$$D_r\|x(t)\| = \lim_{h\downarrow 0} \frac{\|x(t+h)\| - \|x(t)\|}{h} = \lim_{h\downarrow 0} \frac{\|x(t)+hx'(t)\| - \|x(t)\|}{h}$$
$$\leq \lim_{h\downarrow 0} \frac{h\|x'(t)\|}{h} = \|x'(t)\|. \qquad \square$$

Exercise 21. Work out a corresponding result in that $D_\ell\|x(t)\|$ exists on $(a, b]$ and $D_\ell\|x(t)\| \leq \|x'(t)\|$ on $(a, b]$. Here, you will take into account that $h < 0$.

Corollary 4.3. *Assume that $\varphi(t, x)$ is continuous and nonnegative valued on an open $D \subseteq \mathbb{R} \times \mathbb{R}$, and let $x_M(t)$ be the maximal solution of $x' = \varphi(t, x)$ satisfying $x(t_0) = x_0 \geq 0$. Let $y(t)$ be a $C^{(1)}$ n-vector valued function on $[t_0, t_0 + a]$ such that $\|y'(t)\| \leq \varphi(t, \|y(t)\|)$ on $[t_0, t_0 + a]$ and $\|y(t_0)\| \leq x_0$. Then $\|y(t)\| \leq x_M(t)$ on any common right interval of existence of $y(t)$ and $x_M(t)$. Further if $z_m(t)$ is the minimal solution of $x' = -\varphi(t, x)$ satisfying $x(t_0) = x_1$, and again, if $y(t)$ is a $C^{(1)}$ n-vector valued function $[t_0, t_0 + a]$ with $\|y'(t)\| \leq \varphi(t, \|y(t)\|)$ on $[t_0, t_0 + a]$ and $\|y(t_0)\| \geq x_1$, then $\|y(t)\| \geq z_m(t)$ on any common right interval of existence of $y(t)$ and $z_m(t)$.*

Note: In this case, $x' = \varphi(t, x)$ is called a *scalar comparison equation*.

Proof. By Lemma 4.1, $D_r\|y(t)\| \leq \|y'(t)\|$, and so by the hypotheses, $D_r\|y(t)\| \leq \varphi(t, \|y(t)\|)$. Now since $\|y(t)\|$ is a real-valued function and $\|y(t_0)\| \leq x_0$, if we apply Theorem 4.1 to $\|y(t)\|$, we have $\|y(t)\| \leq x_M(t)$.

For the other part, consider $\|y(t+h)\| - \|y(t)\| \geq -\|y(t+h) - y(t)\|$. For $h > 0$,

$$\frac{\|y(t+h)\| - \|y(t)\|}{h} \geq - \left\| \frac{y(t+h) - y(t)}{h} \right\|.$$

Hence, $D_r\|y(t)\| \geq -\|y'(t)\|$, and so by the hypothesis, $D_r\|y(t)\| \geq -\varphi(t, \|y(t)\|)$. Applying one of the corresponding 3 parts of Theorem 4.1 (in Exercise 19), and using $\|y(t)\|$ as the real-valued function, we have $\|y(t)\| \geq z_m(t)$. $\qquad \square$

Corollary 4.4. *Let $\varphi(t,x)$ be continuous on $[t_0, t_0 + a] \times \mathbb{R}$ and be nondecreasing in x, for each fixed t. Let $x_M(t)$ be the maximal solution of the IVP*

$$\begin{cases} x' = \varphi(t,x), \\ x(t_0) = x_0, \end{cases}$$

and let $v(t)$ satisfy $v(t) \le x_1 + \int_{t_0}^t \varphi(s, v(s))\, ds$ on $[t_0, t_0 + a]$, where $x_1 \le x_0$. Then $v(t) \le x_M(t)$ on the intersection of the right maximal internal for $x_M(t)$ with $[t_0, t_0 + a]$.

Proof. Set $z(t) \equiv x_1 + \int_{t_0}^t \varphi(s, v(s))\, ds$ on $[t_0, t_0 + a]$. Then $v(t) \le z(t)$ on $[t_0, t_0 + a]$. Also $z'(t) = \varphi(t, v(t))$. Since φ is nondecreasing, then $z'(t) = \varphi(t, v(t)) \le \varphi(t, z(t))$. Moreover $z(t_0) = x_1 \le x_0$, and hence by Theorem 4.1, $z(t) \le x_M(t)$. Thus $v(t) \le x_M(t)$. $\qquad\square$

Theorem 4.2. *Let $f(t,y)$ be continuous on a slab $[t_0, t_0 + a] \times \mathbb{R}^n$ and assume $\|f(t,y)\| \le \varphi(t, \|y\|)$, where $\varphi(t,x)$ is continuous on $[t_0, t_0 + a] \times [0, +\infty)$ with $\varphi(t,x) \ge 0$. Let $y(t)$ be a solution of the IVP*

$$\begin{cases} y' = f(t,y), \\ y(t_0) = y_0. \end{cases} \tag{4.1}$$

Then, if $x_M(t)$ is the maximal solution of the IVP

$$\begin{cases} x' = \varphi(t,x), \\ x(t_0) = \|y_0\|, \end{cases} \tag{4.2}$$

and if $x_M(t)$ exists on $[t_0, t_0 + a]$, it follows that $y(t)$ extends to $[t_0, t_0 + a]$.

Proof. Assume that $y(t)$ is a solution of (4.1) which does not extend to $[t_0, t_0 + a]$. Then the maximal internal of existence for $y(t)$ is of the form $[t_0, t_0 + \eta)$ with $0 < \eta \le a$ and $\|y(t)\| \to +\infty$ as $t \to (t_0 + \eta)^-$. However, on $[t_0, t_0 + \eta)$,

$$D_r \|y(t)\| \le \|y'(t)\| = \|f(t, y(t))\| \le \varphi(t, \|y(t)\|).$$

By Theorem 4.1, $\|y(t)\| \le x_M(t)$ on $[t_0, t_0 + \eta]$. But $x_M(t)$ exists on $[t_0, t_0 + a]$; thus, there exists $\widehat{M} > 0$ such that $x_M(t) \le \widehat{M}$ on $[t_0, t_0 + a]$, which contradicts $\|y(t)\| \to +\infty$ as $t \to (t_0 + \eta)^-$.

Therefore if y is a solution of (4.1), then $y(t)$ extends to $[t_0, t_0 + a]$. $\quad\square$

Remark 4.2 (Application). Consider the IVP for the linear system.

$$\begin{cases} y' = A(t)y + h(t) = f(t, y), \\ y(t_0) = y_0, \end{cases} \tag{4.3}$$

where $A(t)$ is a continuous $n \times n$ matrix function on an interval $I \subseteq \mathbb{R}$, $h(t)$ is a continuous n-vector function on I, and $(t_0, y_0) \in I \times \mathbb{R}^n$.

By the continuity of A and h, f is Lipschitz on each subset $[a, b] \times \mathbb{R}^n$, $[a, b] \subseteq I$, and so by Theorem 3.3, IVP's for (4.3) have unique solutions on I. Now $\|f(t, y)\| \leq \|A(t)\|\|y(t)\| + \|h(t)\|$ on $I \times \mathbb{R}^n$, and so if we let $\varphi(t, x) = \|A(t)\|x + \|h(t)\|$, then a scalar comparison equation as in Theorem 4.2 for an extension to the right is given by

$$x' = \|A(t)\|x + \|h(t)\|.$$

Exercise 22. Calculate the unique solution of the nonhomogeneous scalar IVP

$$\begin{cases} x' = \|A(t)\|x + \|h(t)\|, \\ x(t_0) = \|y_0\|, \end{cases}$$

which will be the maximal solution, since solutions of IVP's are unique. (This is a simple differential equation which can be solved using integrating factors.) Conclude that the solution $y(t)$ of (4.3) satisfies $\|y(t)\| \leq x_M(t)$ on $[t_0, \infty) \cap I$. This implies the existence of an extension of $y(t)$ and also establishes a bound on $\|y(t)\|$.

Exercise 23. There is another way in which you can determine a bound on $\| y(t) \|$. The solution $y(t)$ of (4.3) is also a solution of

$$y(t) = y_0 + \int_{t_0}^{t} (A(s)y(s) + h(s))\, ds.$$

Hence,

$$\|y(t)\| \leq \|y_0\| + \int_{t_0}^{t} (\|A(s)\|\|y(s)\| + \|h(s)\|)\, ds, \quad \forall t \geq t_0 \text{ in } I.$$

Apply Theorem 1.2 (Gronwall inequality) to the inequality to obtain a bound on $\|y(t)\|$ on $[t_0, \infty) \cap I$ and compare your result with the one given in Exercise 22. Does Corollary 4.4 of Theorem 4.1 apply?

4.2 Kamke Uniqueness Theorem

One of the principal uses of Theorems 4.1 and 4.2 and the corollaries is to obtain uniqueness theorems.

Theorem 4.3 (Kamke Uniqueness Theorem). *Let $f(t,y)$ be continuous on $Q = \{(t,y) \mid t_0 \leq t \leq t_0 + a, \|y - y_0\| \leq b\}$. Let $\varphi(t,u)$ be a real-valued function satisfying the following conditions:*

(1) *φ is continuous on $(t_0, t_0 + a] \times [0, 2b]$.*

(2) *$\varphi(t,0) \equiv 0$ on $(t_0, t_0 + a]$.*

(3) *For any $0 < \varepsilon \leq a$, $u(t) \equiv 0$ is the only solution of $u' = \varphi(t,u)$ on $(t_0, t_0 + \varepsilon]$ which satisfies $u(t) \to 0$ and $\frac{u(t)}{t - t_0} \to 0$ as $t \downarrow t_0$.*

Assume that for any $(t, y_1), (t, y_2) \in Q$ with $t > t_0$,

$$\|f(t, y_1) - f(t, y_2)\| \leq \varphi(t, \|y_1 - y_2\|).$$

Then the IVP

$$\begin{cases} y' = f(t,y), \\ y(t_0) = y_0 \end{cases}$$

has only one solution on any interval $[t_0, t_0 + \varepsilon]$, with $0 < \varepsilon \leq a$.

Proof. Assume the conclusion of the theorem is false so that the IVP $y' = f(t,y)$, $y(t_0) = y_0$ has distinct solutions $y_1(t)$ and $y_2(t)$ on $[t_0, t_0 + \varepsilon]$ with $0 < \varepsilon \leq a$. Let $y(t) \equiv y_1(t) - y_2(t)$. Then there exists a $t_1 \in (t_0, t_0 + \varepsilon]$ such that $\|y(t_1)\| = \|y_1(t_1) - y_2(t_1)\| > 0$ and $\|y(t)\| < 2b$ on $[t_0, t_1]$. Then on $(t_0, t_1]$,

$$D_\ell \|y(t)\| \leq \|y'(t)\| = \|y_1'(t) - y_2'(t)\| = \|f(t, y_1(t)) - f(t, y_2(t))\|$$

$$\leq \varphi(t, \|y_1(t) - y_2(t)\|) = \varphi(t, \|y(t)\|).$$

Now let $v_m(t)$ be the minimal solution of the IVP,

$$\begin{cases} u' = \varphi(t,u), \\ u(t_1) = \|y(t_1)\|, \end{cases}$$

and let $(\alpha, t_1]$ be the left maximal interval of existence for $v_m(t)$ (of course, $(\alpha, t_1] \subseteq (t_0, t_1]$).

It follows from part (3) of Exercise 19 that $v_m(t) \leq \|y(t)\|$ on $(\alpha, t_1]$. We claim that $v_m(t)$ can be continued to $(t_0, t_1]$ such that $0 \leq v_m(t) \leq \|y(t)\|$ (not necessarily as a minimal solution all the way to t_0, but as a nonnegative solution nevertheless).

First, if $t_0 < \alpha < t_1$ and there exists α' with $\alpha < \alpha' < t_1$, such that $v_m(t) > 0$ on $(\alpha', t_1]$ and $v_m(\alpha') = 0$, then by continuity, $v'_m(\alpha') = \varphi(\alpha', v_m(\alpha')) = \varphi(\alpha', 0) = 0$. So,

$$\widehat{v}_m(t) = \begin{cases} 0, & t_0 < t < \alpha', \\ v_m(t), & \alpha' \le t \le t_1, \end{cases}$$

is a solution of $u' = \varphi(t, u)$ and satisfies $0 \le \widehat{v}_m(t) \le \|y(t)\|$ on $(t_0, t_1]$.

For the other case, if $t_0 < \alpha < t_1$ and $v_m(t) > 0$ on $(\alpha, t_1]$, since $\varphi(t, u)$ is bounded and continuous on $[\alpha, t_1] \times [0, 2b]$, it follows from our results of continuation of solutions that $v_m(t)$ can be continued to a solution on $[\alpha, t_1]$.

(i) If $v_m(\alpha) = 0$, repeat the argument of the previous case with $\alpha' = \alpha$.

(ii) If $v_m(\alpha) > 0$, and of course $0 < v_m(\alpha) \le \|y(\alpha)\|$, then $v_m(t)$ could be continued yet further to the **left** as **a minimal solution** of $u' = \varphi(t, u)$ (with $u(\alpha) = v_m(\alpha)$), and satisfy $0 \le v_m(t) \le \|y(t)\|$ on $(\alpha - \delta, t_1]$, for some $\delta > 0$, and hence as such is still a minimal solution of $u' = \varphi(t, u)$, $u(t_1) = \|y(t_1)\|$, which is a contradiction to the fact that $(\alpha, t_1]$ is left maximal.

Therefore, either by construction, or from the impossibility of case (ii) above, it follow that $0 \le v_m(t) \le \|y(t)\|$ on $(t_0, t_1]$. Hence

$$0 \le v_m(t) \le \|y_1(t) - y_2(t)\|.$$

So, $v_m(t) \to 0$, as $t \to t_0$. Also,

$$0 \le \frac{v_m(t)}{t - t_0} \le \left\| \frac{y_1(t) - y_2(t)}{t - t_0} \right\|, \quad \text{for } t_0 < t \le t_1,$$

and so as $t \downarrow t_0$, we have

$$0 \le \frac{v_m(t)}{t - t_0} \le \left\| \frac{y_1(t) - y_2(t)}{t - t_0} \right\|$$
$$= \left\| \frac{y_1(t) - y_1(t_0)}{t - t_0} - \frac{y_2(t) - y_2(t_0)}{t - t_0} \right\| \to \|y'_1(t_0) - y'_2(t_0)\| = 0,$$

i.e., $v_m(t), \frac{v_m(t)}{t - t_0} \to 0$, as $t \downarrow t_0$. From condition (3), $v_m(t) \equiv 0$ on $(t_0, t_1]$; this is a contradiction to $v_m(t_1) = \|y(t_1)\| > 0$.

Therefore, it follows that $y_1(t)$ and $y_2(t)$ are not distinct solutions of the IVP. $\qquad \square$

Corollary 4.5 (Nagumo). *If $f(t, y)$ is continuous on $Q = \{(t, y) \,|\, t_0 \le t \le t_0 + a, \|y - y_0\| \le b\}$ and if for any points $(t, y_1), (t, y_2) \in Q$ with $t > t_0$,*

$\|f(t, y_1) - f(t, y_2)\| \leq \frac{\|y_1 - y_2\|}{t - t_0}$, *then the solution of the IVP*

$$\begin{cases} y' = f(t, y), \\ y(t_0) = y_0, \end{cases}$$

is unique to the right.

Proof. Define $\varphi(t, u) = \frac{u}{t - t_0}$ on $(t_0, t_0 + a] \times [0, 2b]$. Then φ satisfies (1) and (2) of Theorem 4.3. Consider $(t_1, u_0) \in (t_0, t_0 + a] \times [0, 2b]$ and the IVP

$$\begin{cases} u' = \frac{u}{t - t_0}, \\ u(t_1) = u_0. \end{cases}$$

It follows that $u(t) = \frac{u_0 (t - t_0)}{t_1 - t_0}$ is the unique solution. It is the case that $u(t) \to 0$ as $t \to t_0$, however, $\frac{u(t)}{t - t_0} = \frac{u_0}{t_1 - t_0} \to 0$ unless $u_0 = 0$. Thus condition (3) of Theorem 4.3 is also satisfied.

Therefore, there is a unique solution to the right by Theorem 4.3. □

Theorem 4.4. *Let $f(t, y)$ be a continuous real-valued function on $Q = [t_0, t_0 + a] \times [y_0 - b, y_0 + b] \subseteq \mathbb{R} \times \mathbb{R}$. Let $\varphi(t, u)$ be a continuous real-valued function on $(t_0, t_0 + a] \times [0, 2b]$ and assume φ is nondecreasing in u for each fixed t. Then, if the hypothesis of Theorem 4.3 are satisfied, it follows that the sequence of Picard iterates $y_0(t) = y_0$, $y_n(t) = y_0 + \int_{t_0}^{t} f(s, y_{n-1}(s)) \, ds$, $n \geq 1$, converges uniformly on $[t_0, t_0 + \alpha]$ to a solution of the IVP*

$$\begin{cases} y' = f(t, y), \\ y(t_0) = y_0, \end{cases}$$

where $\alpha = \min\left\{a, \frac{b}{M}\right\}$, $M = \max_Q |f(t, y)|$.

We note that Theorem 4.4 applies to first order scalar equations.

Chapter 5

Linear Systems of Differential Equations

5.1 Linear Systems of Differential Equations

We now turn to a detailed discussion of linear systems:

(1) $x' = A(t)x$, a homogeneous first order system; and
(2) $x' = A(t)x + f(t)$, a nonhomogeneous first order system.

We will assume that $A(t) = (a_{ij}(t))$ is a continuous $n \times n$ matrix function on an interval I and that the entries $a_{ij}(t)$ are real- or complex-valued. We will also assume that $f(t)$ is a continuous n-vector function on I with real or complex values.

We have previously shown that, if $\|x\| = \max_{1 \le i \le n}|x_i|$, for $x \in \mathbb{R}^n$, then for A a constant matrix,

$$\|A\| = \sup_{\|x\|=1} \|Ax\| = \max_{1 \le i \le n}\left(\sum_{j=1}^{n}|a_{ij}|\right).$$

Now, we want to discuss the induced norm $\|A\|$, where $\|x\| = (\sum_{i=1}^{n}|x_i|^2)^{\frac{1}{2}}$ is the Euclidean norm of $\mathbb{R}$. Prior to this, we will discuss some properties of the inner product. An inner product is a mapping $\langle \cdot, \cdot \rangle : \mathbb{C}^n \times \mathbb{C}^n \to \mathbb{C}$, where $\mathbb{C}$ is the set of complex numbers, defined by:

$$\langle x, y \rangle \equiv \sum_{j=1}^{n} x_j \, \overline{y}_j, \quad \text{where } x, y \in \mathbb{C}^n.$$

Then

(1) $\langle \alpha x + \beta z, y \rangle = \sum_{j=1}^{n} (\alpha x_j + \beta z_j)\overline{y}_j = \alpha\langle x, y \rangle + \beta\langle z,y \rangle$ (linear in the first component).

(2) $\langle x, \alpha y + \beta z \rangle = \overline{\alpha} \langle x, y \rangle + \overline{\beta} \langle x, z \rangle$ (conjugate linear in the second component).

(3) $\langle x, x \rangle = \sum_{j=1}^{n} |x_j|^2 = \|x\|^2$. So, $\langle x, x \rangle \geq 0$ and $\langle x, x \rangle = 0$ iff $x = 0$.

(4) $\langle x, y \rangle = \overline{\langle y, x \rangle}$.

Lemma 5.1. *Let* $x = (x_1, x_2, \ldots, x_n)$, $y = (y_1, y_2, \ldots, y_n) \in \mathbb{C}^n$. *Then*

$$|\langle x, y \rangle| = \left| \sum_{j=1}^{n} x_j \overline{y}_j \right| \leq \sum_{j=1}^{n} |x_j| \cdot |y_j| \leq \sqrt{\sum_{j=1}^{n} |x_j|^2} \cdot \sqrt{\sum_{j=1}^{n} |y_j|^2} = \|x\|\|y\|.$$

Proof. The first part is true by the triangle inequality and after that, the Schwartz inequality applies.   □

Thus, if $A = (a_{ij})$ and $x \in \mathbb{R}^n$, $\|x\| = \left(\sum_{i=1}^{n} |x_i|^2 \right)^{\frac{1}{2}}$, then

$$Ax = \left[\sum_{j=1}^{n} a_{1j} x_j, \sum_{j=1}^{n} a_{2j} x_j, \ldots, \sum_{j=1}^{n} a_{nj} x_j \right]^T,$$

so that

$$\|Ax\| = \left[\sum_{i=1}^{n} \left| \sum_{j=1}^{n} a_{ij} x_j \right|^2 \right]^{\frac{1}{2}}$$

$$\leq \left[\sum_{i=1}^{n} \left(\sum_{j=1}^{n} |a_{ij}|^2 \right) \left(\sum_{j=1}^{n} |x_j|^2 \right) \right]^{\frac{1}{2}} \quad \text{(By Schwartz inequality)}$$

$$= \sqrt{\sum_{i=1}^{n} \left(\sum_{j=1}^{n} |a_{ij}|^2 \right) \|x\|^2} = \sqrt{\sum_{i,j=1}^{n} |a_{ij}|^2} \|x\|.$$

Hence,

$$\|A\| \leq \sqrt{\sum_{i,j=1}^{n} |a_{ij}|^2}.$$

This expression will not equal $\|A\|$ unless the rows and columns of A are scalar multiples of each other. In a later setting, we may make use of this expression as an upper bound for $\|A\|$.

Let's now let B be a normed vector space and suppose $h : [a, b] \subseteq \mathbb{R} \to B$ is continuous at t_0, i.e., $\lim_{t \to t_0} \|h(t) - h(t_0)\| = 0$.

Now if we take $\|x\| = \max_{1 \leq i \leq n} |x_i|$ so that $\|A\| = \max_{1 \leq i \leq n} \left(\sum_{j=1}^{n} |a_{ij}| \right)$, then clearly $\|x\| \leq \sum_{i=1}^{n} |x_i|$ and $\|x\| \geq |x_{i_0}|$, for any fixed $1 \leq i_0 \leq n$.

Also, $|a_{pq}| \leq \|A\| \leq \sum_{i,j=1}^{n} |a_{ij}|$, for any $1 \leq p, q \leq n$. So, if $A(t) = (a_{ij}(t))$, then

$$|a_{pq}(t) - a_{pq}(t_0)| \leq \|A(t) - A(t_0)\| \leq \sum_{i,j=1}^{n} |a_{ij}(t) - a_{ij}(t_0)|.$$

Theorem 5.1. $A(t)$ *is continuous at t_0 iff $a_{pq}(t)$ is continuous at t_0, for all $1 \leq p, q \leq n$.*

We now state and prove some basic theorems concerning linear systems.

Lemma 5.2. *Assume $A(t)$ and $f(t)$ are continuous on $I \subseteq \mathbb{R}$. Then for each point $(t_0, x_0) \in I \times \mathbb{R}^n$, each of the IVP's*

$$\begin{cases} x' = A(t)\,x, \\ x(t_0) = x_0, \end{cases} \tag{5.1}$$

$$\begin{cases} x' = A(t)\,x + f(t), \\ x(t_0) = x_0, \end{cases} \tag{5.2}$$

has a unique solution on I.

Proof. Since $f(t) \equiv 0$ in (5.1), we will directly deal with (5.2). First, for $x_1, x_2 \in \mathbb{R}^n$, we have

$$\|A(t)x_1 + f(t) - A(t)x_2 - f(t)\| \leq \|A(t)\| \cdot \|x_1 - x_2\|.$$

Thus, if $[a, b] \subseteq I$ and $K = \max_{t \in [a,b]} \|A(t)\|$, we have

$$\|A(t)\,x_1 + f(t) - A(t)\,x_2 - f(t)\| \leq K \|x_1 - x_2\|,$$

so that a Lipschitz condition is satisfied wrt x on each compact $[a, b] \subseteq I$. By Theorem 1.3, each of the IVP's has a unique solution on I. $\square$

Corollary 5.1. *If $A(t)$ is continuous on $I \subseteq \mathbb{R}$, then for each $t_0 \in I$, the unique solution of*

$$\begin{cases} x' = A(t)\,x, \\ x(t_0) = 0, \end{cases}$$

is $x(t) \equiv 0$.

Our next results are concerned with the solution space of (5.1).

Theorem 5.2. *Assume that $A(t)$ and $f(t)$ are continuous on $I \subseteq \mathbb{R}$. Let $x_1(t), x_2(t), \ldots, x_m(t)$ be solutions of (5.1); let $\alpha_1, \alpha_2, \ldots, \alpha_m$ be constants. Then $x(t) = \sum_{j=1}^{m} \alpha_j x_j(t)$ is a solution of (5.1). Moreover, if $y_1(t)$ and $y_2(t)$ are solution of (5.2), then $y_1(t) - y_2(t)$ is a solution of (5.1).*

Proof. First, by

$$x'(t) = \sum_{j=1}^{m} \alpha_j x_j'(t) = \sum_{j=1}^{m} \alpha_j A(t) x_j(t) = A(t) \sum_{j=1}^{m} \alpha_j x_j(t) = A(t) \, x(t),$$

we have $x(t) = \sum_{j=1}^{m} \alpha_j x_j(t)$ is a solution of (5.1).

For the other part,

$$\begin{aligned}
y_1'(t) - y_2'(t) &= A(t)y_1(t) + f(t) - (A(t)y_2(t) + f(t)) \\
&= A(t)\,(y_1(t) - y_2(t))\,.
\end{aligned}$$

Therefore, $y_1(t) - y_2(t)$ is a solution of (5.1).   □

We will denote the set of all continuous $\mathbb{R}^n$-valued functions on an interval $I \subseteq \mathbb{R}$ by $C[I, \mathbb{R}^n]$. Similarly, $C[I, \mathbb{C}^n]$ is the set of all continuous $\mathbb{C}^n$-valued functions on $I \subseteq \mathbb{R}$. $C[I, \mathbb{R}^n]$ is a vector space over $\mathbb{R}$ and $C[I, \mathbb{C}^n]$ is a vector space over $\mathbb{C}$.

Theorem 5.2 shows that the solution space (collection of solutions) of $x' = A(t)x$ is a subspace of $C[I, \mathbb{R}^n]$ or $C[I, \mathbb{C}^n]$ depending upon whether $A(t)$ is real- or complex-valued.

Recall that if V is a vector space over a field K, then a subset $U \subseteq V$ is said to be *linearly independent* (L.I.) in case for any finite set of distinct vectors $\{x_1, x_2, \ldots, x_m\}$ in U, $\sum_{j=1}^{m} \alpha_j x_j = 0$ iff $\alpha_1 = \alpha_2 = \cdots = \alpha_m = 0$, where $\alpha_1, \ldots, \alpha_m \in K$. A subset $W \subseteq V$ is said to *span* V in case every $x \in V$ can be expressed as a finite linear combination of vectors from W. A subset $B \subseteq V$ is said to be *a basis* for V in case B is L.I. and B spans V. All bases of a given vector space V have the same cardinality.

In particular, if $V = \mathbb{R}^n$, $K = \mathbb{R}$, then $\dim V = n$ with basis $\{e_i\}_{i=1}^{n}$. In fact, any set of n L.I. vectors in $\mathbb{R}^n$ is a basis.

Theorem 5.3. *Let $x_1(t), x_2(t), \ldots, x_m(t)$ be solutions of (5.1). Then, if there are constants $\alpha_1, \alpha_2, \ldots, \alpha_m$ and a point $t_0 \in I$ such that $\sum_{j=1}^{m} \alpha_j x_j(t_0) = 0$, then it follow that $\sum_{j=1}^{m} \alpha_j x_j(t) \equiv 0$ on I.*

Consequently, the solution space of (5.1) *is an n-dimensional subspace of $C(I, \mathbb{R}^n)$ (or $C[I, \mathbb{C}^n]$). Furthermore, if $x_1(t), x_2(t), \ldots, x_n(t)$ are solutions of* (5.1) *such that, for some $t_0 \in I$, $x_1(t_0), \ldots, x_n(t_0)$ are L.I. in $\mathbb{R}^n$ (or $\mathbb{C}^n$), then $x_1(t), \ldots, x_n(t)$ constitute a basis for the solution space of* (5.1).

Proof. For the first assertion, assume $x_1(t), \ldots, x_m(t)$ are solutions of (5.1) and that $\alpha_1, \ldots, \alpha_m$ are scalars and $t_0 \in I$ such that $\sum_{j=1}^{m} \alpha_j x_j(t_0) = 0$. By Theorem 5.2, $x(t) = \sum_{j=1}^{m} \alpha_j x_j(t)$ is a solution of (5.1). Moreover, $x(t_0) = 0$, and so by Corollary 5.1, $x(t) \equiv 0$ on I.

Assume now that $x_1(t), \ldots, x_n(t)$ are solutions of (5.1) such that at some t_0, $x_1(t_0), \ldots, x_n(t_0)$ are L.I. vectors in $\mathbb{R}^n$. Such solutions exist; e.g., for each $1 \leq j \leq n$, let $x_j(t)$ be the solution of

$$\begin{cases} x' = A(t)x, \\ x(t_0) = e_j. \end{cases}$$

We claim that any such set $x_1(t), \ldots, x_n(t)$ are L.I. vectors in $\mathbb{R}^n$, for all points $t \in I$. If not, then there are scalars $\alpha_1, \ldots, \alpha_n$ not all zero and $t_1 \in I$, such that $\sum_{j=1}^{n} \alpha_j x_j(t_1) = 0$. By the first assertion, $\sum_{j=1}^{n} \alpha_j x_j(t) \equiv 0$, and in particular $\sum_{j=1}^{n} \alpha_j x_j(t_0) = 0$, which contradicts to the L.I. of $\{x_j(t_0)\}_{j=1}^{n}$. Hence $\alpha_j = 0$, $1 \leq j \leq n$, and $x_1(t), \ldots, x_n(t)$ are L.I. in $C[I, \mathbb{R}^n]$,

We now show that $x_1(t), \ldots, x_n(t)$ span the solution space of (5.1). Let $z(t)$ be any solution of (5.1). Since $x_1(t_0), \ldots, x_n(t_0)$ are L.I. in $\mathbb{R}^n$, they must constitute a basis for $\mathbb{R}^n$. Now $z(t_0) \in \mathbb{R}^n$, so there are scalars $\alpha_1, \ldots, \alpha_n$ such that $z(t_0) = \sum_{j=1}^{n} \alpha_j x_j(t_0)$. Now $z(t)$ and $\sum_{j=1}^{n} \alpha_j x_j(t)$ are both solutions of

$$\begin{cases} x' = A(t)x, \\ x(t_0) = z(t_0), \end{cases}$$

and so by Lemma 5.2, $z(t) \equiv \sum_{j=1}^{n} \alpha_j x_j(t)$, for $t \in I$.

Therefore, $\{x_j(t)\}_{j=1}^{n}$ span and hence form a basis for the solution space of (5.1). $\qquad\square$

Corollary 5.2. *The solution set of* (5.2) *consists of all* $y(t) = y_0(t) + \sum_{j=1}^n \alpha_j x_j(t)$, *where* $y_0(t)$ *is some fixed solution of* (5.2) *and* $x_1(t), \ldots, x_n(t)$ *are L.I. solutions of* (5.1).

Proof. Consider the IVP (5.2):

$$\begin{cases} x' = A(t)x + f(t), \\ x(t_0) = c = (c_1, \ldots, c_n)^T. \end{cases}$$

Let $y(t)$ be the solution.

Let $x_j(t)$ satisfy

$$\begin{cases} x' = A(t)x, \\ x(t_0) = e_j, \end{cases}$$

for each $1 \le j \le n$. Then, these x_j's are L.I.

Let $y_0(t)$ be the solution of

$$\begin{cases} x' = A(t)x + f(t), \\ x(t_0) = 0. \end{cases}$$

Then the solution of (5.2) will be $y(t) = y_0(t) + \sum_{j=1}^n c_j x_j(t)$, since $y(t) - y_0(t)$ and $\sum_{j=1}^n c_j x_j(t)$ satisfy the same IVP for (5.1). $\qquad\square$

Exercise 24. Assume that $f(t, x)$ is continuous and has continuous first partial derivatives wrt the components of x on an open set $D \subseteq \mathbb{R} \times \mathbb{R}^n$. Let $(t_0, x_0) \in D$ and (α, ω) be the maximal interval of existence of $x(t; t_0, x_0)$, the solution of $x' = f(t, x)$, $x(t_0) = x_0$. Let $y(t) = \frac{\partial x(t; t_0, x_0)}{\partial t_0}$ and $x_j(t) = \frac{\partial x(t; t_0, x_0)}{\partial x_{0j}}$, $1 \le j \le n$. Show $y(t) = \sum_{j=1}^n (-f_j(t_0, x_0)) x_j(t)$ on (α, ω), where $f_j(t, x)$ is the jth component of $f(t, x)$; i.e.,

$$f(t, x) = \begin{bmatrix} f_1(t, x) \\ f_2(t, x) \\ \vdots \\ f_n(t, x) \end{bmatrix}.$$

5.2 Some Properties of Matrices

Let $\mathcal{M}_n$ denote the set of all $n \times n$ matrices with complex entries. Then, we have the following properties of $\mathcal{M}_n$.

(1) $\mathcal{M}_n$ is a vector space over $\mathbb{C}$ with vector operations defined for $\alpha \in \mathbb{C}$, $A = (a_{ij})$ and $B = (b_{ij})$ as

$$A + B = (a_{ij} + b_{ij}), \quad \alpha A = (\alpha a_{ij}).$$

As a complex vector space, $\dim \mathcal{M}_n = n^2$.

(2) Define $AB = C$, where $c_{ij} = \sum_{k=1}^{n} a_{ik} b_{kj}$. In general, $AB \neq BA$. We have associative properties, distributive, $\alpha(AB) = A(\alpha B) = (\alpha A)B$, etc. $\mathcal{M}_n$ is an "algebra".

(3) $A^T = (a_{ij})^T = (a_{ji})$ and so $(AB)^T = B^T A^T$. Also, $\overline{A} = (\overline{a}_{ij})$ and the adjoint, $A^* = (\overline{A})^T = \overline{(A^T)}$. The identity matrix $I = (\delta_{ij})$, and if associated with A, there is a matrix B such that $AB = BA = I$, then A is said to be *nonsingular*. We write $A^{-1} = B$.

If $A \in \mathcal{M}_n$ is such that $\det A \neq 0$, and $B = (b_{ij})$, where b_{ij} is the cofactor of a_{ij}, then A is nonsingular and $A^{-1} = \frac{1}{\det A} B$.

$\boxed{\text{Exercise}}$ **25.** Prove if A is nonsingular, then $\det A \neq 0$. (Note: $\det(AB) = \det A \cdot \det B$).

It is also the case that A is nonsingular iff A is a one-to-one transformation. Equivalently, the fact that A is nonsingular is equivalent to the statement: $Ax = 0$ iff $x = 0$.

Other facts which may be established are

$$\left. \begin{array}{l} \det A = \det A^T \\ \det \overline{A} = \overline{\det A} \end{array} \right\} \implies \det A^* = \overline{\det A}.$$

From this, A is nonsingular iff $\overline{A}$, A^T, A^* are nonsingular.

If A, B are nonsingular, then so is AB and $(AB)^{-1} = B^{-1}A^{-1}$. Also, $AA^{-1} = I$ implies $(AA^{-1})^* = I^* = I$, and so $(A^{-1})^* A^* = I$. Thus, $(A^{-1})^* = (A^*)^{-1}$. Similarly $(\overline{A})^{-1} = \overline{(A^{-1})}$ and $(A^T)^{-1} = (A^{-1})^T$.

(4) Range and null space of $A \in \mathcal{M}_n$.

The range space $\mathcal{R}(A) = \{Ax \,|\, x \in \mathbb{C}^n\}$, and the null space $\mathcal{N}(A) = \{x \in \mathbb{C}^n \,|\, Ax = 0\}$. From liner algebra, $\dim(\mathcal{R}(A)) + \dim(\mathcal{N}(A)) = \dim(\mathbb{C}^n) = n$. Moreover,

$$A \text{ is nonsingular} \quad \text{iff} \quad \mathcal{N}(A) = \{0\} \text{ and } \mathcal{R}(A) = \mathbb{C}^n;$$
$$\text{iff} \quad \text{the column vectors of } A \text{ are L.I.;}$$
$$\text{iff} \quad \text{the row vectors of } A \text{ are L.I.}$$

Let us recall that the inner product of $x, y \in \mathbb{C}^n$ is given by $\langle x, y \rangle = \sum_{i=1}^{n} x_i \overline{y}_i$. From this, we have $\langle Ax, y \rangle = \langle x, A^*y \rangle$ and consequently, for a given $A \in \mathcal{M}_n$ and a vector $b \in \mathbb{C}^n$, $Ax = b$ has a solution $x \in \mathbb{C}^n$ iff $\langle b, y \rangle = 0$, for all $y \in \mathcal{N}(A^*)$ and $\dim \mathcal{R}(A) = \dim \mathcal{R}(A^*)$ and $\dim \mathcal{N}(A) = \dim \mathcal{N}(A^*)$.

(5) Properties of $\|A\|$.

  (i) $\|Ax\| \geq 0$ and $\|Ax\| = 0$ iff $A = (0)$.
  (ii) $\|Ax\| \leq \|A\| \|x\|$.
  (iii) $\|\alpha A\| = \|\alpha\| \|A\|$.
  (iv) $\|A + B\| \leq \|A\| + \|B\|$.
  (v) $\|AB\| \leq \|A\| \|B\|$.

Proof of (iv). We use $\|A\| = \inf \{M > 0 \mid \|Ax\| \leq M \|x\|$, for all $x \in \mathbb{C}^n\}$. So, for all $x \in \mathbb{C}^n$,

$$\|(A + B)x\| = \|Ax + Bx\| \leq \|Ax\| + \|Bx\|$$
$$\leq \|A\| \|x\| + \|B\| \|x\| = (\|A\| + \|B\|) \|x\|.$$

If $\|A\| + \|B\| = M' > 0$, then $\|(A + B)x\| \leq M' \|x\|$, for all $x \in \mathbb{C}^n$. Hence,

$$\|A + B\| \leq \|A\| + \|B\|.$$

$\square$

Proof of (v). Notice $\|(AB)x\| = \|A(Bx)\| \leq \|A\| \|Bx\| \leq \|A\| \|B\| \|x\|$, for any x. If $M' = \|A\| \|B\|$, then $\|(AB)x\| \leq M' \|x\|$, and so $\|AB\| \leq \|A\| \|B\|$. $\square$

We continue with this briefest of a review with glance concerning metric spaces.

A *metric space* (S, d) consists of a set S and a mapping $d : S \times S \to \mathbb{R}^+$ such that for any $s_1, s_2, s_3 \in S$,

(a) $d(s_1, s_2) \geq 0$, and $d(s_1, s_2) = 0$ iff $s_1 = s_2$;
(b) $d(s_1, s_2) = d(s_2, s_1)$;
(c) $d(s_1, s_2) \leq d(s_1, s_3) + d(s_3, s_2)$.

A metric space S is said to be *complete* in case every Cauchy sequence in S converges in S.

For $A = (a_{ij}), B = (b_{ij}) \in \mathcal{M}_n$, define $d(A, B) = \|A - B\|$. Then d is a metric on $\mathcal{M}_n$. From statements above, we have

$$|a_{pq} - b_{pq}| \leq \|A - B\| \leq \sum_{i,j=1}^{n} |a_{ij} - b_{ij}|.$$

Remark 5.1. $\mathcal{M}_n$ is a complete metric space.

Proof. Let $\{A_k\}$ be a Cauchy sequence in $\mathcal{M}_n$. Then for each $\varepsilon > 0$, there exists N_ε such that $d(A_k, A_l) = \|A_k - A_l\| < \varepsilon$, for all $k, l \geq N_\varepsilon$. Thus, for all $1 \leq p, q \leq n$,

$$\left| a_{pq}^{(k)} - a_{pq}^{(l)} \right| \leq \|A_k - A_l\| < \varepsilon, \text{ for all } k, l \geq N_\varepsilon, \text{ where } A_k = \left(a_{ij}^{(k)} \right).$$

Hence, $\left\{ a_{pq}^{(k)} \right\}$ is a Cauchy sequence in $\mathbb{C}$, for each fixed $1 \leq p, q \leq n$. Since $\mathbb{C}$ is complete, for each $1 \leq p, q \leq n$, there exists an $a_{pq}^{(0)}$ such that $\lim_{k \to \infty} a_{pq}^{(k)} = a_{pq}^{(0)}$. If we let $A_0 = \left(a_{ij}^{(0)} \right)$, then $\|A_k - A_0\| \leq \sum_{i,j=1}^{n} \left| a_{ij}^{(k)} - a_{ij}^{(0)} \right| \to 0$, as $k \to +\infty$.

Hence, $\lim_{k \to \infty} d(A_k, A_0) = 0$, and consequently, $\mathcal{M}_n$ is complete in the metric $d(A, B) = \|A - B\|$. $\square$

Since $\mathcal{M}_n$ is also a normal vector space, $\mathcal{M}_n$ is a Banach space. In fact, $\mathcal{M}_n$ is a Banach Algebra.

Remark 5.2. In a complete metric space, a sequence converges iff it is Cauchy.

Exercise 26. Let $\{A_k\}_{k=1}^{\infty}$ be a sequence in $\mathcal{M}_n$. Assume $\lim_{k \to \infty} A_k = A_0$ and that $A_k C = C A_k$, for all $k \geq 1$, where $C \in \mathcal{M}_n$. Prove that $A_0 C = C A_0$.

5.3 Infinite Series of Matrices and Matrix-Valued Functions

For what is to follow concerning linear differential equations, we need to discuss what is meant by infinite series of matrices.

With an infinite series of matrices: $\sum_{k=0}^{\infty} \alpha_k A^k$, $A^0 = I$, we can associate the sequence of partial sums, $S_n = \sum_{k=0}^{n} \alpha_k A^k$.

If $S = \sum_{k=0}^{\infty} \alpha_k A^k$, then the sequence of partial sums converges and $\lim_{n \to \infty} S_n = S$. If $\{S_k\}$ diverges, then $\sum_{k=0}^{\infty} \alpha_k A^k$ diverges.

Example 5.1. Consider

$$\sum_{k=0}^{\infty} \frac{t^k A^k}{k!} = I + tA + \frac{t^2 A^2}{2!} + \cdots.$$

This series converges for any $A \in \mathcal{M}_n$ and $t \in \mathbb{R}$.
To see this convergence, consider the series of numbers

$$\sum_{k=0}^{\infty} \frac{|t|^k \|A\|^k}{k!},$$

which converges to $e^{|t| \|A\|}$, for all $|t|, \|A\|$. Thus, for this series of numbers, the Cauchy criterion is satisfied by its sequence of partial sums, $\{S_n'\}$. Hence, for each $\varepsilon > 0$, there exists $N_\varepsilon > 0$ such that

$$\|S_n' - S_m'\| = \sum_{k=m+1}^{n} \frac{|t|^k \|A\|^k}{k!} < \varepsilon, \text{ for all } n, m \geq N_\varepsilon.$$

Now let $\{S_n\}$ be the sequence of partial sums associated with $\sum_{k=0}^{\infty} \frac{t^k A^k}{k!}$. Then for the same $\varepsilon > 0$ and for all $n > m \geq N_\varepsilon$,

$$\|S_n - S_m\| = \left\| \sum_{k=m+1}^{n} \frac{t^k A^k}{k!} \right\| \leq \sum_{k=m+1}^{n} \frac{|t|^k \|A\|^k}{k!} < \varepsilon.$$

Hence, $\{S_n\}$ form a Cauchy sequence in $\mathcal{M}_n$ and hence $\{S_n\}$ converges by the completeness of $\mathcal{M}_n$; i.e., $\sum_{k=0}^{\infty} \frac{t^k A^k}{k!}$ converges, for all t, A.

Since the above series closely resembles the exponential series, we define $e^{tA} \equiv \sum_{k=0}^{\infty} \frac{t^k A^k}{k!}$.

In a similar way we can show that $\sum_{k=0}^{\infty} \frac{(-1)^k t^{2k} A^{2k}}{(2k)!}$ converges; and we define $\cos(tA) \equiv \sum_{k=0}^{\infty} \frac{(-1)^k t^{2k} A^{2k}}{(2k)!}$. Other recognizable series are treated similarly.

In what follows, we define what we mean by continuity, differentiability, and integrability of a matrix-valued function. Let $A : [a, b] \to \mathcal{M}_n$.

1. $A(t)$ is continuous at $t_0 \in [a, b]$, in case $\lim_{t \to t_0} \|A(t) - A(t_0)\| = 0$.
2. $A(t)$ has derivative $B \in \mathcal{M}_n$ at $t = t_0$, in case

$$\lim_{h \to 0} \left\| \frac{A(t_0 + h) - A(t_0)}{h} - B \right\| = 0.$$

We write $A'(t_0) = B$.
3. Let $[a, b] \subseteq \mathbb{R}$ and $\Pi : a = t_0 < t_1 < \cdots < t_n = b$ be a partition of $[a, b]$ with $\|\Pi\| = \max_{1 \leq i \leq n} \Delta t_i$. Select $\tau_i \in [t_{i-1}, t_i]$. We say that $A(t)$ is

Riemann integrable on $[a, b]$ in case there exists a $B \in \mathcal{M}_n$ such that for all $\varepsilon > 0$, there exists $\delta > 0$ such that

$$\left\| \sum_{i=1}^{n} A(\tau_i) \Delta t_i - B \right\| < \varepsilon,$$

for any partition Π with $\|\Pi\| < \delta$.

We can easily prove the following:

(1) $A(t)$ is continuous at t_0 iff each $a_{ij}(t)$ is continuous at t_0.

Proof. The conclusion is proved by making use of

$$|a_{pq}(t) - a_{pq}(t_0)| \leq \|A(t) - A(t_0)\| \leq \sum_{i,j=1}^{n} |a_{ij}(t) - a_{ij}(t_0)|. \qquad \square$$

(2) $A(t)$ is differentiable at t_0 iff $a_{ij}(t)$ is differentiable at t_0.

Proof. It is proved by making use of

$$\left| \frac{a_{pq}(t_0 + h) - a_{pq}(t_0)}{h} - b_{pq} \right| \leq \left\| \frac{A(t_0 + h) - A(t_0)}{h} - B \right\|$$

$$\leq \sum_{i,j=1}^{n} \left| \frac{a_{ij}(t_0 + h) - a_{ij}(t_0)}{h} - b_{ij} \right|.$$

This says, in fact, that $A'(t_0) = B = (b_{ij}) = \left(a'_{ij}(t_0) \right)$. $\qquad \square$

(3) $A(t)$ is integrable iff $a_{ij}(t)$ is integrable.

Proof. It can be proved by making use of the above definition of integrability of $A(t)$ and inequalities similar to those in (1) and (2). It will follow that

$$\int_a^b A(t)\, dt = \left(\int_a^b a_{ij}(t)\, dt \right). \qquad \square$$

Now if we assume differentiability, we can verify the following:

(4) If $A(t), B(t) \in \mathcal{M}_n$, then $(A(t)B(t))' = A'(t)B(t) + A(t)B'(t)$.

(5) If $A(t) \in \mathcal{M}_n$, $x(t), y(t) \in \mathbb{C}^n$, then $[A(t)x(t)]' = A'(t)x(t) + A(t)x'(t)$, and $\langle x(t), y(t) \rangle' = \langle x'(t), y(t) \rangle + \langle x(t), y'(t) \rangle$.

(6) If $\alpha(t) \in \mathbb{C}$, $A(t) \in \mathcal{M}_n$, $x(t) \in \mathbb{C}^n$, then $[\alpha(t)A(t)]' = \alpha'(t)A(t) + \alpha(t)A'(t)$, and $(\alpha(t)x(t))' = \alpha'(t)x(t) + \alpha(t)x'(t)$.

(7) If $A(t) \in \mathcal{M}_n$ and $A(t)$ is nonsingular, then $A^{-1}(t)$ is differentiable and

$$[A^{-1}(t)]' = A^{-1}(t)A'(t)A^{-1}(t).$$

Proof. First $A^{-1}(t) = \frac{1}{\det A(t)} B(t)$, where $B(t)$ is the transposed matrix of cofactors of A. Since $A(t) = (a_{ij}(t))$ is differentiable, it follows that $b_{ij}(t)$'s are differentiable and also that $\det A(t)$ is differentiable. Hence, $A^{-1}(t)$ is differentiable.

Now

$$0 = I' = \left[A(t) A^{-1}(t) \right]' = A'(t) \left(A^{-1}(t) \right) + A(t) \left[A^{-1}(t) \right]'$$

yields

$$A(t) \left[A^{-1}(t) \right]' = -A'(t) A^{-1}(t),$$

which implies

$$\left[A^{-1}(t) \right]' = -A^{-1}(t) A'(t) A^{-1}(t). \qquad \square$$

(8) $\overline{[A(t)]}' = \overline{A'(t)}$, $[A^T(t)]' = [A'(t)]^T$, $[A^*(t)]' = [A'(t)]^*$.

(9) Suppose the matrix $\Phi(t)$ satisfies the D.E. $\Phi'(t) = A(t)\Phi(t)$. Then,

$$[\Phi^*(t)]' = [\Phi'(t)]^* = [A(t)\Phi(t)]^* = \Phi^*(t)A^*(t).$$

If Φ is nonsingular, then $[\Phi^{-1}(t)]' = -\Phi^{-1}(t)\,\Phi'(t)\,\Phi^{-1}(t)$. Therefore,

$$\begin{aligned}
[(\Phi^{-1}(t))^*]' &= -[\Phi^{-1}(t)]^* [\Phi'(t)]^* [\Phi^{-1}(t)]^* \\
&= -[\Phi^*(t)]^{-1} [\Phi^*(t)]' [\Phi^{-1}(t)]^* \\
&= -[\Phi^*(t)]^{-1}\Phi^*(t)A^*(t)[\Phi^{-1}(t)]^* \\
&= -A^*(t)[\Phi^{-1}(t)]^*.
\end{aligned}$$

Thus $\Phi^{-1}(t)^*$ satisfies the D.E.

$$\psi' = -A^*(t)\psi. \tag{5.3}$$

5.4   Linear Matrix System

Now from our previous theory, since $h(t, x) = A(t)x + f(t)$, where $A(t) \in \mathcal{M}_n$ is continuous and $f(t) \in \mathbb{C}^n$ is continuous, $h(t, x)$ satisfies a Lipschitz condition wrt x on each slab $[a, b] \times \mathbb{C}^n$, it follows from the Picard Existence Theorem that the linear system

$$\begin{cases} x' = A(t)x + f(t), \\ x(t_0) = c, \end{cases} \tag{5.4}$$

where $t_0 \in J$ and J is an interval, has a unique solution on all of J which is a solution of the integral equation

$$x(t) = c + \int_{t_0}^t [A(s)x(s) + f(s)]ds, \quad \text{for all } t \in J \tag{5.5}$$

and is also the uniform limit of Picard iterates on each compact subinterval of J.

In summary, if $A(t)$ and $f(t)$ are continuous on J, then the unique solution of the IVP (5.4) can be written as

$$x(t) = \sum_{j=1}^{n} c_j x_j(t) + z(t),$$

where $c = (c_1, \ldots, c_n)^T$, $x_j(t)$ is the solution of the IVP

$$\begin{cases} x' = A(t)\, x, \\ x(t_0) = e_j, \end{cases}$$

and $z(t)$ is the solution of

$$\begin{cases} x' = A(t)x + f(t), \\ x(t_0) = 0. \end{cases}$$

In a similar way, we can take under our consideration IVP's for a matrix system,

$$\begin{cases} \Phi'(t) = A(t)\Phi(t) + B(t), \\ \Phi(t_0) = C, \end{cases} \tag{5.6}$$

where $A(t)$ and $B(t)$ are continuous $n \times n$ matrix functions on an interval $J \subseteq \mathbb{R}$, and C is a fixed constant $n \times n$ matrix, and $t_0 \in J$.

The matrix IVP (5.6) is equivalent to the IVP for a system of n^2 scalar linear equations

$$\begin{cases} \phi'_{ij} = \sum_{k=1}^{n} a_{ik}(t)\phi_{kj} + b_{ij}(t), & 1 \le i, j \le n, \\ \phi_{ij}(t_0) = c_{ij}, & 1 \le i, j \le n. \end{cases}$$

Also, $\Phi(t) = (\phi_{ij}(t))$ is a solution of the matrix equation iff each of its column vectors $\phi_j(t) = [\phi_{1j}(t), \ldots, \phi_{nj}(t)]^T$, $1 \le j \le n$ is a solution of the vector equation

$$\phi'_j(t) = A(t)\, \phi_j(t) + b_j(t),$$

where $b_j(t)$ is the jth column of $B(t)$. (Note: One obtains the jth column of $A(t)\Phi(t)$ by multiplying $A(t)$ by the jth column of $\Phi(t)$; i.e., $A(t)\phi_j(t)$.)

In light of our previous discussions, the unique solution of the IVP

$$\begin{cases} \Phi' = A(t)\Phi + B(t), \\ \Phi(t_0) = C \end{cases}$$

can be obtained as the uniform limit of the sequence of Picard iterates as before:

$$\Phi_0(t) = C,$$

$$\Phi_n(t) = C + \int_{t_0}^{t} [A(s)\,\Phi_{n-1}(s) + B(s)]\,ds, \quad n \geq 1.$$

In the case when $A \in \mathcal{M}_n$ is constant, we give special attention to IVP's for the homogeneous matrix equation. Consider

$$\begin{cases} \Phi' = A\Phi, & A \text{ is constant,} \\ \Phi(t_0) = I. \end{cases}$$

By Picard iteration,

$$\Phi_0(t) = I,$$

$$\Phi_1(t) = I + \int_{t_0}^{t} A\,ds = I + (t - t_0)A,$$

$$\Phi_2(t) = I + \int_{t_0}^{t} A[I + (s - t_0)A]\,ds$$

$$= I + A(t - t_0) + \frac{(t - t_0)^2}{2!}A^2,$$

$$\vdots$$

$$\Phi_n(t) = I + (t - t_0)A + \frac{(t - t_0)^2}{2!}A^2 + \cdots + \frac{(t - t_0)^n}{n!}A^n.$$

By previous results, the Picard iterates converge uniformly on *any* compact interval $[a, b] \subseteq \mathbb{R}$, and moreover, the Φ_n consists of the nth partial sum whose limit we designated by the exponential function; i.e., the solution $\Phi(t) = \lim_{n \to \infty} \Phi_n(t) = e^{(t-t_0)A}$.

Observe that since this is a solution of the D.E. above, we also have

$$\frac{d}{dt}e^{(t-t_0)A} = Ae^{(t-t_0)A}.$$

Exercise **27.**

(i) Show that $e^{(t-t_0)A}$ is nonsingular, for each $t \in \mathbb{R}$, and that $\left[e^{(t-t_0)A}\right]^{-1} = e^{-(t-t_0)A}$. (Note: Verifying the equality will demonstrate that $e^{(t-t_0)A}$ is nonsingular.).

(ii) Show $e^{tA}B = Be^{tA}$, for all $t \in \mathbb{R}$, iff $AB = BA$.

(iii) Show $e^{t(A+B)} = e^{tA}e^{tB}$, for all $t \in \mathbb{R}$, iff $AB = BA$.

Note: e^{tB} is the unique solution of

$$\begin{cases} X' = BX, \\ X(0) = I. \end{cases} \tag{5.7}$$

From this point on, if we haven't already stated it, assume that matrix and vector function are continuous on an internal $J \subseteq \mathbb{R}$.

Lemma 5.3. *If $X(t)$ is a matrix solution of the matrix equation $X' = A(t)X$ and if $\mathbf{c} \in \mathbb{C}^n$, then $x(t) = X(t)\mathbf{c}$ is a solution of the vector system $x' = A(t)x$.*

Proof. Notice

$$\frac{d}{dt}[X(t)\mathbf{c}] = X'(t)\mathbf{c} = A(t)X(t)\mathbf{c} = A(t)\left(X(t)\mathbf{c}\right).$$

Thus, if $x(t) = X(t)\mathbf{c}$, then we have $x'(t) = A(t)x(t)$. $\qquad\square$

Lemma 5.4. *If $X(t)$ is a solution of the matrix equation $X' = A(t)X$, then either $X(t)$ is nonsingular for all $t \in J$, or $X(t)$ is singular for all $t \in J$.*

Proof. Assume that for some $t_0 \in J$, $X(t_0)$ is singular. Now recall that a matrix $B \in \mathcal{M}_n$ is singular iff there exists $\mathbf{c} \in \mathbb{C}^n$, $\mathbf{c} \neq 0$, such that $B\mathbf{c} = 0$. Thus there exists $\mathbf{c} \neq 0$ such that $X(t_0)\mathbf{c} = 0$.

Let $x(t) = X(t)\mathbf{c}$. Then from above, $x(t)$ is a solution of

$$\begin{cases} x' = A(t)x, \\ x(t_0) = X(t_0)\mathbf{c} = 0. \end{cases}$$

By uniqueness of solutions of IVP's, $x(t)$ is the unique solution and hence, $x(t) \equiv 0$, i.e., $X(t)\mathbf{c} \equiv 0$ on J. Hence, $X(t)$ is singular, for all $t \in J$. $\qquad\square$

Note: Lemma 5.4 says that if $\mathbf{c} \in \mathcal{N}(X(t_0))$, for some t_0, then $\mathbf{c} \in \mathcal{N}(X(t))$, for all $t \in J$. This is strong and is due to the fact that $X(t)$ is a solution of the D.E.

Theorem 5.4. *Let $X(t)$ be a solution of $X' = A(t)X$ and let $t_0 \in J$. Then $\det X(t) = \det X(t_0)e^{\int_{t_0}^t \operatorname{Tr} A(s)\, ds}$, where $\operatorname{Tr} A(s) = \sum_{i=1}^n \alpha_{ii}(s)$. (Note: This also confirms Lemma 5.4.)*

Proof. Recall that if $X(t) = (x_{ij}(t))$ is a solution of $X' = A(t)X$, then $x'_{ij}(t) = \sum_{k=1}^{n} a_{i_k} x_{k_j}(t)$, for all $1 \le i, j \le n$.

Now, denoting

$$\det X(t) = \begin{vmatrix} x_{11} & x_{12} & \cdots & x_{1n} \\ x_{21} & x_{22} & \cdots & x_{2n} \\ \vdots & \vdots & \ddots & \vdots \\ x_{n1} & x_{n2} & \cdots & x_{nn} \end{vmatrix},$$

then

$$\frac{d}{dt}[\det X(t)]$$

$$= \begin{vmatrix} x'_{11} & x'_{12} & \cdots & x'_{1n} \\ x_{21} & x_{22} & \cdots & x_{2n} \\ \vdots & \vdots & \ddots & \vdots \\ x_{n1} & x_{n2} & \cdots & x_{nn} \end{vmatrix} + \begin{vmatrix} x_{11} & x_{12} & \cdots & x_{1n} \\ x'_{21} & x'_{22} & \cdots & x'_{2n} \\ \vdots & \vdots & \ddots & \vdots \\ x_{n1} & x_{n2} & \cdots & x_{nn} \end{vmatrix} + \cdots + \begin{vmatrix} x_{11} & x_{12} & \cdots & x_{1n} \\ x_{21} & x_{22} & \cdots & x_{2n} \\ \vdots & \vdots & \ddots & \vdots \\ x'_{n1} & x'_{n2} & \cdots & x'_{nn} \end{vmatrix}.$$

If we select the kth summand

$$\begin{vmatrix} x_{11} & x_{12} & \cdots & x_{1n} \\ \vdots & \vdots & \vdots & \vdots \\ x'_{k1} & x'_{k2} & \cdots & x'_{kn} \\ \vdots & \vdots & \vdots & \vdots \\ x_{n1} & x_{n2} & \cdots & x_{nn} \end{vmatrix},$$

then, since $x'_{kp} = \sum_{j=1}^{n} a_{kj} x_{jp}$, we have

$$\begin{vmatrix} x_{11} & x_{12} & \cdots & x_{1n} \\ \vdots & \vdots & \ddots & \vdots \\ x'_{k1} & x'_{k2} & \cdots & x'_{kn} \\ \vdots & \vdots & \ddots & \vdots \\ x_{n1} & x_{n2} & \cdots & x_{nn} \end{vmatrix} = \begin{vmatrix} x_{11} & x_{12} & \cdots & x_{1n} \\ \vdots & \vdots & \ddots & \vdots \\ \sum_{j=1}^{n} a_{kj} x_{j1} & \sum_{j=1}^{n} a_{kj} x_{j2} & \cdots & \sum_{j=1}^{n} a_{kj} x_{jn} \\ \vdots & \vdots & \ddots & \vdots \\ x_{n1} & x_{n2} & \cdots & x_{nn} \end{vmatrix}$$

$$= a_{kk}(t) \det X(t).$$

(By elementary row operations, every row cancels with each of the sums in the kth row.) Hence,

$$\frac{d}{dt}[\det X(t)] = \sum_{k=1}^{n} a_{kk}(t) \det X(t) = \operatorname{Tr} A(t) \det X(t).$$

If we set $\psi(t) = \det X(t)$, we have then

$$\psi'(t) - \operatorname{Tr} A(t)\psi(t) = 0.$$

Using integrating factors, $e^{-\int_{t_0}^{t} \operatorname{Tr} A(s)ds}\psi(t) = \text{constant} = K$.
So, $K = \det X(t_0)$ and hence,

$$e^{-\int_{t_0}^{t} \operatorname{Tr} A(s)\,ds}\psi(t) = \det X(t_0), \quad \text{or}$$

$$\det X(t) = \det X(t_0)e^{\int_{t_0}^{t} \operatorname{Tr} A(s)\,ds}. \qquad \square$$

Consider now Theorem 5.4 applied to the nth order linear equation,

$$x^{(n)} + p_1(t)x^{(n-1)} + \cdots + p_n(t)x = 0. \qquad (5.8)$$

As a vector system, we write

$$
\begin{aligned}
y_1 &= x, \\
y_2 &= y_1', \\
&\;\vdots \\
y_n &= y_{n-1}',
\end{aligned}
\qquad \implies \qquad
-p_n(t)y_1 - \cdots - p_1(t)y_n = y_n',
$$

or

$$y' = A(t)y, \text{ where } A(t) = \begin{bmatrix} 0 & 1 & 0 & \cdots & 0 \\ 0 & 0 & 1 & \cdots & 0 \\ 0 & 0 & 0 & \cdots & 1 \\ -p_n & -p_{n-1} & -p_{n-2} & \cdots & -p_1 \end{bmatrix}.$$

Furthermore, as we have seen, $y(t)$ is a solution of $y' = A(t)y$ iff $y(t) = (y_1(t), y_1'(t), \ldots, y_1^{(n-1)}(t))^T$, where $y_1(t)$ is a solution of the nth order linear equation (5.8). Thus, $X(t)$ is a matrix solution of $X' = A(t)X$ iff each column of $X(t)$ is of the form of $y(t)$ above; i.e., iff

$$X(t) = \begin{bmatrix} x_1(t) & x_2(t) & \cdots & x_n(t) \\ x_1'(t) & x_2'(t) & & x_n'(t) \\ \vdots & \vdots & \ddots & \vdots \\ x_1^{(n-1)}(t) & x_2^{(n-1)}(t) & \cdots & x_n^{(n-1)}(t) \end{bmatrix},$$

where $x_i(t)$, $1 \le i \le n$ is a solution of the n order linear equation (5.8). Then consider

$$\det X(t) = \begin{vmatrix} x_1(t) & x_2(t) & \cdots & x_n(t) \\ x_1'(t) & x_2'(t) & \cdots & x_n'(t) \\ \vdots & \vdots & \ddots & \vdots \\ x_1^{(n-1)}(t) & x_2^{(n-1)}(t) & \cdots & x_n^{(n-1)}(t) \end{vmatrix},$$

which is called the *Wronskian* of the solutions $x_1(t), \ldots, x_n(t)$ of the nth order linear equation (5.8) and is denoted by $W(t; x_1, \cdots, x_n)$. From Theorem 5.4,

$$W(t; x_1, \cdots, x_n) = W(t_0; x_1, \cdots, x_n)e^{-\int_{t_0}^t p_1(s)\, ds}.$$

Definition 5.1. A nonsingular matrix solution $X(t)$ of the matrix D.E. $X' = A(t)X$ will be called a *fundamental matrix solution*. The fundamental matrix solution of the IVP:

$$\begin{cases} X' = A(t)X, \\ x(t_0) = I, \end{cases}$$

will be denoted by $X(t, t_0)$ (i.e., this is the solution which satisfies $X(t_0, t_0) = I$).

Remark 5.3. The solution of the IVP

$$\begin{cases} X' = A(t)X, \\ X(t_0) = C \end{cases}$$

is given by $X(t, t_0)C$.

Proof. If $X(t)$ is a solution of $X' = A(t)X$ and if $C \in \mathcal{M}_n$, then

$$\frac{d}{dt}[X(t)C] = X'(t)C = A(t)X(t)C = A(t)[X(t)C];$$

i.e., $X(t)C$ is also a solution of the matrix equation. Hence $X(t, t_0)C$ is a solution which satisfies the initial condition $X(t, t_0)C = IC = C$. □

Another observation is, if $Y(t)$ is a solution of $X' = A(t)X$ and $t_0 \in J$, then $Y(t) \equiv X(t, t_0)Y(t_0)$ by the above remark.

Definition 5.2. The system $x' = -A^*(t)x$ is called the *adjoint system* wrt the system $x' = A(t)x$.

| Exercise | 28. (i) Show that for any $s, t, t_0 \in J$, $X(t, t_0) = X(t, s)X(s, t_0)$. Hence $[X(s, t_0)]^{-1} = X(t_0, s)$.

(ii) Let $A(t)$ be continuous on $[0, +\infty)$ and assume that $\|x(t)\|$ is bound on $[0, +\infty)$ for each solution of the vector equation $x' = A(t)x$. Let $X(t)$ be a fundamental matrix solution of $X' = A(t)X$. Then prove that $\|X^{-1}(t)\|$ is bound on $[0, +\infty)$ iff $\text{Re}\{\int_0^t \text{Tr}A(s)ds\}$ is bound below on $[0, +\infty)$.

(iii) Let $X(t)$ be the fundamental matrix solution of $X' = A(t)X$. Then $Y(t)$ is a fundamental matrix solution of the adjoint system $X' = -A^*(t)X$ iff $Y^*(t)X(t) \equiv C$, where is C is nonsingular.

Now let's consider the IVP

$$\begin{cases} x' = A(t)x + f(t), \\ x(t_0) = c \in \mathbb{C}^n. \end{cases}$$

The solution, as we have seen, can be expressed as $x(t) = \sum_{j=1}^n c_j x_j(t) + z(t)$, where $c = [c_1, \ldots, c_n]^T$, $x_j(t)$ is the solution of $x' = A(t)x$, $x(t_0) = e_j$ and $z(t)$ is the solution of $x' = A(t)x + f(t)$, $x(t_0) = 0$. We would like to analyze $z(t)$ a little closer. To begin with, since $X(t, t_0) = [x_1(t), \ldots, x_n(t)]$, we can write the solution $x(t)$ as $x(t) = X(t, t_0)c + z(t)$.

Theorem 5.5 (Variation of Constants Formula). *Let $X(t)$ be a fundamental matrix solution of the matrix D.E. $X' = A(t)X$. Then $z(t) = X(t)\int_{t_0}^t X^{-1}(s)f(s)\, ds$ is the solution of the IVP*

$$\begin{cases} x' = A(t)x + f(t), \\ x(t_0) = 0. \end{cases}$$

Proof. We seek a solution $z(t)$ of $x' = A(t)x + f(t)$ of the form $z(t) = X(t)y(t)$, and we try to determine $y(t)$. Now $\frac{d}{dt}z(t) = X'(t)y(t) + X(t)y'(t)$. Thus, if $z(t)$ is a solution, we must have

$$X'(t)y(t) + X(t)y'(t) = A(t)X(t)y(t) + f(t),$$

which yields

$$A(t)X(t)y(t) + X(t)y'(t) = A(t)X(t)y(t) + f(t),$$

that is, $y'(t) = X^{-1}(t)f(t)$.

Also, to have $z(t_0) = X(t_0)y(t_0) = 0$, we must have $y(t_0) = 0$, since $X(t_0)$ is nonsingular.

Consequently,

$$y(s)\big|_{t_0}^t = \int_{t_0}^t y'(s)\, ds = \int_{t_0}^t X^{-1}(s)f(s)\, ds,$$

and so

$$y(t) = \int_{t_0}^t X^{-1}(s)f(s)\, ds.$$

Since $z(t) = X(t)y(t)$, we conclude $z(t) = X(t)\int_{t_0}^t X^{-1}(s)f(s)\, ds$. $\qquad\square$

Note: $X(t)X^{-1}(t_0)$ is a solution of the IVP

$$\begin{cases} X' = A(t)X, \\ X(t_0) = I, \end{cases}$$

so that by uniqueness, $X(t, t_0) = X(t)X^{-1}(t_0)$. It follows from Theorem 5.5 that

$$z(t) = \int_{t_0}^{t} X(t)X^{-1}(s)f(s)\,ds = \int_{t_0}^{t} X(t,s)f(s)\,ds.$$

Hence, finally the solution of the IVP

$$\begin{cases} x' = A(t)x + f(t), \\ x(t_0) = c \end{cases}$$

can be written as

$$x(t) = X(t,t_0)c + \int_{t_0}^{t} X(t,s)f(s)\,ds.$$

Exercise **29.** Show that the solution of the IVP

$$\begin{cases} x' = A(t)x + f(t), \\ x(t_0) = 0 \end{cases}$$

can also be expressed in the form $x(t) = [Y^*(t)]^{-1} \int_{t_0}^{t} Y^*(s)f(s)\,ds$, where $Y(t)$ is a fundamental matrix solution of $X' = -A^*(t)X$.

Remark 5.4. The unique solution of $X' = A(t)X + B(t), X(t_0) = C$ is $X(t) = X(t,t_0)C + \int_{t_0}^{t} X(t,s)B(s)\,ds$.

5.5 Higher Order Differential Equations

Let us again consider the two equations:

$$x^{(n)} + p_1(t)x^{(n-1)} + \cdots + p_n(t)x = 0 \quad \text{(Homogeneous)}, \tag{5.9}$$

$$x^{(n)} + p_1(t)x^{(n-1)} + \cdots + p_n(t)x = h(t) \quad \text{(Nonhomogeneous)}. \tag{5.10}$$

By our previous results, the equivalent vector systems are

$$y' = A(t)y, \tag{5.11}$$

$$y' = A(t)y + f(t), \tag{5.12}$$

where

$$A(t) = \begin{bmatrix} 0 & 1 & 0 & \cdots & 0 \\ 0 & 0 & 1 & \cdots & 0 \\ \vdots & \vdots & \vdots & \ddots & \vdots \\ 0 & 0 & 0 & \cdots & 1 \\ -p_n & -p_{n-1} & -p_{n-2} & \cdots & -p_1 \end{bmatrix} \text{ and } f(t) = \begin{bmatrix} 0 \\ 0 \\ \vdots \\ h(t) \end{bmatrix}.$$

Let $x_1(t), \ldots, x_m(t)$ be solutions of (5.9) on J. Then (5.11) has corresponding solutions, $y^1(t), \ldots, y^m(t)$ given by

$$y^1(t) = (x_1(t), x_1'(t), \ldots, x_1^{(n-1)}(t))^T,$$
$$\vdots$$
$$y^m(t) = (x_m(t), x_m'(t), \ldots, x_m^{(n-1)}(t))^T.$$

Lemma 5.5. $x_1(t), \ldots, x_m(t)$ *are L.I. in the vector space* $C^{(n-1)}[J, \mathbb{C}]$ *iff* $y^1(t), \ldots, y^m(t)$ *are L.I. in the vector space* $C[J, \mathbb{C}^n]$.

Proof. Assume that $y^1, \ldots, y^m$ are L.D. in $C[J, \mathbb{C}^n]$. So, there exist $\alpha_1, \ldots, \alpha_m$, not all zero, in $\mathbb{C}$ such that $\sum_{j=1}^m \alpha_j y^j(t) \equiv 0$ on J. Hence, the sum in the first component is zero. In other words, $\alpha_1 x_1(t) + \cdots + \alpha_m x_m(t) \equiv 0$ on J. Therefore, $x_1, \ldots, x_m$ are L.D.

Conversely, assume that $x_1, \ldots, x_m$ are L.D. in $C^{(n-1)}[J, \mathbb{C}]$. Then, there exist $\alpha_1, \ldots, \alpha_m$, not all zero, in $\mathbb{C}$ such that $\alpha_1 x_1(t) + \cdots + \alpha_m x_m(t) \equiv 0$ on J. Upon differentiating $n-1$ times, we have $\alpha_1 x_1^{(i)}(t) + \cdots + \alpha_m x_m^{(i)}(t) \equiv 0$ on J, for all $1 \le i \le n-1$. From the manner in which the $y^i(t)$ were constructed, we have $\alpha_1 y^1(t) + \cdots + \alpha_m y^m(t) \equiv 0$ on J, or $y^1, \ldots, y^m$ are L.D.

Now a matrix solution $X(t)$ of (5.11) has the form

$$X(t) = \begin{bmatrix} x_1(t) & x_2(t) & \cdots & x_n(t) \\ x_1'(t) & x_2'(t) & \cdots & x_n'(t) \\ \vdots & \vdots & \ddots & \vdots \\ x_1^{(n-1)}(t) & x_2^{(n-1)}(t) & \cdots & x_n^{(n-1)}(t) \end{bmatrix}.$$

So,

$X(t)$ is a fundamental
matrix solution   iff $X(t)$ is nonsingular,

         iff $X(t)\mathbf{c} = 0$ yields $\mathbf{c} = 0 \in \mathbb{C}$,

         iff   the columns of $X(t)$ are L.I. in $C[J, \mathbb{C}^n]$,

         iff $x_1(t), \ldots, x_n(t)$ are L.I. in $C^{(n-1)}[J, \mathbb{C}]$.

In fact, the fundamental matrix solution $X(t)$ of (5.11) satisfying the initial condition $X(t_0) = I$ has as the jth element in the first row, the solution $x_j(t)$ of (5.9) which satisfies

$$x_j^{(i)}(t_0) = 0, \quad 0 \le i \le n - 1, i \ne j - 1,$$

$$x_j^{(j-1)}(t_0) = 1.$$

Thus, if $x_1, \ldots, x_n$ satisfy the above initial conditions, then by Theorem 5.4,

$$W(t; x_1, \ldots, x_n) = W(t_0; x_1, \ldots, x_n)e^{-\int_{t_0}^t p_1(s)\, ds} = e^{-\int_{t_0}^t p_1(s)\, ds}. \qquad \square$$

Example 5.2. Let $x_1(t)$ and $x_2(t)$ be solution of $x'' + 2tx' - t^2 x = 0$ satisfying the respective initial conditions

$$\begin{cases} x_1(0) = 1, \\ x_1'(0) = 0, \end{cases} \quad \begin{cases} x_2(0) = 0, \\ x_2'(0) = 1. \end{cases}$$

Then

$$\begin{aligned} W(t; x_1, x_2) &= x_1(t)x_2'(t) - x_1'(t)x_2(t) \\ &= W(0; x_1, x_2)e^{-\int_0^t 2s\, ds} \\ &= 1 \cdot e^{-t^2} = e^{-t^2}. \end{aligned}$$

Now the unique solution of the IVP corresponding to (5.9)

$$\begin{cases} y' = A(t)y + f(t), \\ y(t_0) = 0, \end{cases}$$

is given by $y(t) = \int_{t_0}^t X(t, s)f(s)\, ds$, where $X(t, s)$ is the solution of

$$\begin{cases} X' = A(t)X, \\ X(s) = X(s, s) = I. \end{cases}$$

Recall

$$f(s) = \begin{bmatrix} 0 \\ 0 \\ \vdots \\ h(s) \end{bmatrix}.$$

Now *the first component* $x(t)$ of the solution $y(t) = \int_{t_0}^{t} X(t, s) f(s) \, ds$ is a solution of equation (5.10). Since $y(t_0) = 0$, $x(t)$ satisfies the initial conditions $x^{(i)}(t_0) = 0$, $0 \leq i \leq n - 1$.

The first component $x(t)$ of $y(t)$ is given by

$$x(t) = \int_{t_0}^{t} u(t, s) h(s) \, ds,$$

where $u(t, s)$ is the solution of the IVP of equation (5.5)

$$\begin{cases} x^{(n)} + p_1(t) x^{(n-1)} + \cdots + p_n(t) x = 0, \\ x^{(i)}(s) = 0, \quad 0 \leq i \leq n - 2, \\ x^{(n-1)}(s) = 1. \end{cases}$$

And $u(t, s)$ is frequently called the *Cauchy Function* for the equation

$$x^{(n)} + p_1(t) x^{(n-1)} + \cdots + p_n(t) x = 0.$$

In summary, the unique solution of the IVP

$$\begin{cases} x^{(n)} + p_1(t) x^{(n-1)} + \cdots + p_n(t) x = h(t), \\ x^{(i)}(t_0) = 0, \quad 0 \leq i \leq n - 1. \end{cases}$$

is given by

$$x(t) = \int_{t_0}^{t} u(t, s) h(s) \, ds.$$

Example 5.3.

(1) $x^{(n)} = 0$.

Then $u(t, s) = \frac{(t-s)^{n-1}}{(n-1)!}$. Thus, the unique solution of

$$\begin{cases} x^{(n)} = h(t), \\ x^{(j)}(t_0) = 0, \ 0 \leq i \leq n - 1, \end{cases}$$

is given by

$$x(t) = \int_{t_0}^{t} \frac{(t-s)^{n-1}}{(n-1)!} h(s) \, ds.$$

(2) $x'' + x = 0$.

Then $u(t, s) = \sin(t - s)$, Thus, the unique solution of

$$\begin{cases} x'' + x = h(t), \\ x(t_0) = x'(t_0) = 0, \end{cases}$$

is given by

$$x(t) = \int_{t_0}^{t} \sin(t - s)h(s)\,ds.$$

(3) From (1), we can derive Taylor's formula with remainder.

Theorem 5.6. *Let $f(t) \in C^{(n-1)}[a, b]$ and suppose $f^{(n)}(t)$ exists and is integrable on $[a, b]$. Then*

$$f(t) \equiv \sum_{k=0}^{n-1} \frac{f^{(k)}(a)(t-a)^k}{k!} + \int_{a}^{t} \frac{(t-s)^{n-1}f^{(n)}(s)}{(n-1)!}\,ds, \ for\ t \in [a, b].$$

Proof. Consider the IVP

$$\begin{cases} x^{(n)}(t) = f^{(n)}(t), \\ x^{(i)}(a) = f^{(i)}(a), \quad 0 \le i \le n - 1. \end{cases}$$

This has a unique solution which must be $f(t)$. The solution of the homogenous equation $x^{(n)} = 0$ is given by a suitable combination of the solutions $x_j(t)$ satisfying

$$x^{(i)}(a) = 0,\ 0 \le i \le n - 1,\ i \ne j - 1, \quad x^{(j-1)}(a) = 1.$$

Note $x_j(t) = \frac{(t-a)^{j-1}}{(j-1)!}$ by (1). Hence

$$x(t) = f(t) = \sum_{j=1}^{n} f^{(j-1)}(a)x_j(t) + \int_{a}^{t} \frac{(t-s)^{n-1}f^{(n)}(s)}{(n-1)!}\,ds. \qquad \square$$

(4) Applying (2) we solve

$$\begin{cases} x'' + x = h(t), \\ x(0) = -1, \quad x'(0) = 2, \end{cases}$$

where

$$h(t) = \begin{cases} 0, & -\infty < t \le \pi, \\ 1, & t > \pi. \end{cases}$$

First, considering the homogenous equation $x'' + x = 0$, two solutions satisfying

$$\begin{cases} x_1(0) = 1, \\ x_1'(0) = 0, \end{cases} \qquad \begin{cases} x_2(0) = 0, \\ x_2'(0) = 1, \end{cases}$$

are given respectively by $x_1(t) = \cos t$, $x_2(t) = \sin t$. Then the solution of the IVP is

$$x(t) = -x_1(t) + 2x_2(t) + \int_0^t \sin(t-s)h(s)\,ds$$

$$= -\cos t + 2\sin t + \int_0^t \sin(t-s)\,\chi_{[\pi,\infty)}(s)\,ds,$$

where $\chi_{[\pi,\infty)}(s)$ is the indicator function.

(5) For this example, we recall our variation of constants formula in Theorem 5.5 for the solution of

$$\begin{cases} x' = A(t)x + f(t), \\ x(t_0) = 0. \end{cases}$$

The solution was given by $z(t) = \int_{t_0}^t X(t,s)f(s)\,ds$, where we arrived at this by looking for a solution of the form $z(t) = X(t)y(t)$ where $X(t)$ is a fundamental matrix solution of the homogeneous equation and $y(t)$ satisfied $X(t)y'(t) = f(t)$ or $y(t) = \int_{t_0}^t X^{-1}(s)f(s)\,ds$.

In light of this recall from elementary differential equation courses, we consider the equation

$$x'' + p_1(t)x' + p_2(t)x = h(t). \tag{5.13}$$

Suppose $x_1(t)$ and $x_2(t)$ are L.I. solutions of the homogeneous equation $x'' + p_1(t)x' + p_2(t)x = 0$. Then we sought (in the elementary differential equation course) a solution of (5.13) in the form $z(t) = c_1(t)x_1(t) + c_2(t)x_2(t)$ (i.e., the "constants" vary, hence the name).

Now if we assume $c_1'(t)x_1(t) + c_2'(t)x_2(t) = 0$, and if $c_1'(t)x_1'(t) + c_2'(t)x_2'(t) = h(t)$, then it is the case that $z(t) = c_1(t)x_1(t) + c_2(t)x_2(t)$ is a solution of (5.13).

Thus, if

$$c_1'(t)x_1(t) + c_2'(t)x_2(t) = 0,$$
$$c_1'(t)x_1'(t) + c_2'(t)x_2'(t) = h(t),$$

then we have a solution of (5.13). Putting this in matrix form

$$\begin{bmatrix} x_1(t) & x_2(t) \\ x_1'(t) & x_2'(t) \end{bmatrix} \cdot \begin{bmatrix} c_1'(t) \\ c_2'(t) \end{bmatrix} = \begin{bmatrix} 0 \\ h(t) \end{bmatrix},$$

which corresponds to $X(t)y'(t) = f(t)$. That is,

$$\begin{bmatrix} c_1(t) \\ c_2(t) \end{bmatrix} = \int_{t_0}^t \begin{bmatrix} x_1(s) & x_2(s) \\ x_1'(s) & x_2'(s) \end{bmatrix}^{-1} \begin{bmatrix} 0 \\ h(s) \end{bmatrix} ds$$

and from this we have a solution of (5.13): $z(t) = c_1 x_1 + c_2 x_2$.

For our next consideration we will be concerned with solutions of the adjoint equation associated with

$$x^{(n)} + p_1(t)x^{(n-1)} + \cdots + p_n(t)x = 0, \qquad (5.14)$$

where we assume the $p_i(t)$'s are complex-valued functions which are continuous on an interval $J \subseteq \mathbb{R}$.

From the associated vector system $y'(t) = A(t)y$, consider the adjoint system $y'(t) = -A^*(t)y$, where

$$-A^*(t) = \begin{bmatrix} 0 & 0 & \dots & 0 & \overline{p}_n \\ -1 & 0 & \dots & 0 & \overline{p}_{n-1} \\ 0 & -1 & \dots & 0 & \overline{p}_{n-2} \\ \vdots & \vdots & \vdots & \vdots & \vdots \\ 0 & 0 & \dots & -1 & \overline{p}_1 \end{bmatrix} \text{ and } y = \begin{bmatrix} y_1 \\ \vdots \\ y_n \end{bmatrix}.$$

In component form, the *adjoint system* becomes

$$\begin{aligned} y_1' &= \overline{p}_n y_n, \\ y_2' &= -y_1 + \overline{p}_{n-1} y_n, \\ y_3' &= -y_2 + \overline{p}_{n-2} y_n, \\ &\vdots \\ y_{n-1}' &= -y_{n-2} + \overline{p}_2 y_n, \\ y_n' &= -y_{n-1} + \overline{p}_1 y_n. \end{aligned}$$

If the p_i's are continuous on J, then any solution $y(t)$ of the adjoint system belongs to $C^{(1)}[J, \mathbb{C}^n]$. To consider a scalar equation adjoint to (5.14), one is usually concerned with differentiability of products $\overline{p}_j y_n$.

If $\overline{p}_1 y_n$ is differentiable, then y_n' is differentiable, since $y_n' = -y_{n-1} + \overline{p}_1 y_n$. So, from above,

$$\begin{aligned} y_n'' &= -y_{n-1}' + (\overline{p}_1 y_n)' \\ &= -(-y_{n-2} + \overline{p}_2 y_n) + (\overline{p}_1 y_n)' \\ &= y_{n-2} - \overline{p}_2 y_n + (\overline{p}_1 y_n)'. \end{aligned}$$

Similarly, if $\overline{p}_2 y_n$ and $(\overline{p}_1 y_n)'$ are in turn differentiable, then y_n''' is differentiable and

$$y_n''' = y_{n-2}' - (\overline{p}_2 y_n)' + (\overline{p}_1 y_n)'',$$
$$\vdots$$

Continuing, we obtain that the nth component y_n satisfies

$$y_n^{(n)}(t) + \sum_{j=1}^{n} (-1)^j \left(\overline{p}_j y_n \right)^{(n-j)} = 0. \qquad (5.15)$$

We conclude that, if $y(t)$ is a solution of the adjoint system on J such that for its nth component $y_n(t)$, we have $\overline{p}_j y_n$ is $(n-j)$th differentiable on J, for all $j = 1, 2, \ldots, n$, then $y_n(t)$ is a solution of the scalar equation (5.15) on J. The equation (5.15) is frequently called the *adjoint* of the original scalar equation (5.14).

In the above development, the differentiation of the p_i's is important. For, if $\overline{p}_1 y_n$ has a derivative at $t_0 \in J$ and if $y_n(t_0) \neq 0$, then $\overline{p}_1$ is differentiable, since $\overline{p}_1 = \frac{\overline{p}_1 y_n}{y_n}$. Hence, if $\overline{p}_1$ is not differentiable, the product $\overline{p}_1 y_n$ is not likely to have a derivative unless $y_n = 0$, which isn't often.

$\boxed{\text{Exercise}}$ **30.** The first part of this exercise is not related to the above theory, but perhaps you will be able to complete it successfully. Assume that $p_1(t), \ldots, p_n(t)$ are continuous on an interval J. Then, if $[a, b] \subseteq J$, any solution $x(t) \not\equiv 0$ of $x^{(n)} + p_1(t)x^{(n-1)} + \cdots + p_n(t)x = 0$ can have at most a finite number of zeros on $[a, b]$. Is this true for the nth component $y_n(t)$ of a solution $y(t) \not\equiv 0$ of the adjoint system? (For this second part, use the above theory.)

$\boxed{\text{Exercise}}$ **31.** Recalling that $\left[X^{-1}(t)\right]^*$ is a solution of the adjoint matrix system $Y' = -A^*(t)Y$, where $X(t)$ is a solution of $X' = A(t)X$, obtain 3 linearly independent solutions of $x''' - (tx)'' + (3t^2 x)' - t^3 x = 0$ (i.e., solutions of $x''' - tx'' + (3t^2 - 2)x' + (6t - t^3)x = 0$) in terms of Wronskians of solutions of the equation which it is the adjoint of, namely $x''' + tx'' + 3t^2 x' + t^3 x = 0$. (Need Wronskians in computing the inverse matrix.)

$\boxed{\text{Exercise}}$ **32.** Prove that $X' = A(t)X$ has a fundamental matrix solution $X(t)$ which is unitary; that is, $X^{-1}(t) = X^*(t)$, iff $A(t) \equiv -A^*(t)$. Is this possible for systems obtained from nth order scalar equations?

$\boxed{\text{Exercise}}$ **33.** If $A(t) \equiv -A^*(t)$ (i.e., $A(t)$ is skew-symmetric), show that for any solution $x(t)$ of $x' = A(t)x$, $\langle x(t), x(t) \rangle$ is a constant.

5.6 Systems of Equations with Constant Coefficient Matrices

For some time now, we will consider systems of equations with constant coefficient matrices. Let A **be a constant matrix.**

Definition 5.3. A number $\lambda \in \mathbb{C}$ is said to be an *eigenvalue* of the matrix $A \in \mathcal{M}_n$ in case there exists a nonzero vector c such that $Ac = \lambda c$, where c is called an *eigenvector*.

Lemma 5.6. *The eigenvalues of A are the roots of the polynomial equation* $\det[A - \lambda I] = 0$.

Proof. Suppose $Ac = \lambda c$, for some nonzero vector c. Then,

$Ac = \lambda c$ iff $(A - \lambda I)c = 0$ iff $[A - \lambda I]$ is singular iff $\det[A - \lambda I] = 0$.

$\square$

Definition 5.4. The $\det[A - \lambda I]$ is called the *characteristic polynomial* of the matrix A.

Lemma 5.7. *If $\lambda_1, \ldots, \lambda_m$ are distinct eigenvalues of A and if $c_1, \ldots, c_m$ are the associated eigenvectors, then the vectors $c_1, \ldots, c_m$ are L.I.*

Proof. Assume that $c_1, \ldots, c_m$ are not L.I. Then, there exists a first integer j with $1 < j < m$ such that $c_1, \ldots, c_{j-1}$ are L.I., but $c_1, \ldots, c_j$ are L.D. Hence, there exist scalars $\alpha_1, \ldots, \alpha_j$, not all zero, such that $\alpha_1 c_1 + \cdots + \alpha_j c_j = 0$. Since $Ac_i = \lambda c_i$, we have, $0 = A \cdot 0 = A(\alpha_1 c_1 + \cdots + \alpha_j c_j) = \alpha_1 \lambda_1 c_1 + \cdots + \alpha_j \lambda_j c_j$. Moreover, $\lambda_j \neq 0$. For if $\lambda_j = 0$, we would have above that $\alpha_1 \lambda_1 c_1 + \cdots + \alpha_{j-1} \lambda_{j-1} c_{j-1} = 0$ and $\lambda_i \neq 0$, for $1 \leq i \leq j - 1$. So, $\alpha_i \lambda_i = 0$, $1 \leq i \leq j-1$, since $c_1, \ldots, c_{j-1}$ are assumed L.I., thus $\alpha_i = 0$, $1 \leq i \leq j-1$. Since $\alpha_1 c_1 + \cdots + \alpha_j c_j = 0$ and $c_j \neq 0$, we would then have $\alpha_j = 0$, which is a contradiction to the assumption that they were not all zero.

Consequently $\lambda_j \neq 0$, and since the λ_i's are all distinct, we have $\frac{\lambda_i}{\lambda_j} \neq 1$, for all $i \neq j$.

Now $\alpha_j c_j = -\frac{1}{\lambda_j}(\alpha_1 \lambda_1 c_1 + \cdots + \alpha_{j-1} \lambda_{j-1} c_{j-1})$ and $\alpha_j c_j = -(\alpha_1 c_1 + \cdots + \alpha_{j-1} c_{j-1})$ yield

$$\left(1 - \frac{\lambda_1}{\lambda_j}\right)\alpha_1 c_1 + \cdots + \left(1 - \frac{\lambda_{j-1}}{\lambda_j}\right)\alpha_{j-1} c_{j-1} = 0.$$

Arguing as above, we have $\alpha_i = 0$, $1 \leq i \leq j$. Again, we get a contradiction, thus our supposition concerning the existence of $\alpha_1, \ldots, \alpha_j$, not all zero, is false and we have that $c_1, \ldots, c_m$ are L.I. $\square$

Lemma 5.8. *If $\lambda_1, \ldots, \lambda_m$ are distinct eigenvalues of A with associated eigenvectors $c_1, \ldots, c_m$, then $x_1(t) = e^{\lambda_1 t} c_1, \ldots, x_m(t) = e^{\lambda_m t} c_n$ are L.I. solutions of $x' = Ax$.*

Proof. First we verify that $x_1(t), \ldots, x_m(t)$ are solutions:

$$\frac{d}{dt} x_j(t) = \frac{d}{dt} e^{\lambda_j t} c_j = \lambda_j e^{\lambda_j t} c_j = e^{\lambda_j t}(\lambda_j c_j) = e^{\lambda_j t}(Ac_j)$$
$$= Ae^{\lambda_j t} c_j = Ax_j(t).$$

Hence, $\{x_j(t)\}_{j=1}^m$ is a set of solutions of $x' = Ax$.
Now $\{x_j(0)\}_{j=1}^m = \{c_j\}_{j=1}^m$ which are L.I. by the above lemma. Thus, by Theorem 5.3, $x_1(t), \ldots, x_m(t)$ are L.I. solutions. $\qquad\square$

In relating this lemma to results from your elementary ODE's course, consider the nth order scalar linear equation $x^{(n)} + p_1 x^{(n-1)} + \cdots + p_n x = 0$, where the p_i's are constants. Then there is the usual associated first order system:

$$y' = Ay = \begin{bmatrix} 0 & 1 & 0 & \cdots & 0 & 0 \\ 0 & 0 & 1 & \cdots & 0 & 0 \\ 0 & 0 & 0 & \cdots & 0 & 0 \\ \vdots & \vdots & \vdots & \ddots & \vdots & \vdots \\ 0 & 0 & 0 & \cdots & 0 & 1 \\ -p_n & -p_{n-1} & -p_{n-2} & \cdots & -p_2 & -p_1 \end{bmatrix} y.$$

Hence,

$$A - \lambda I = \begin{bmatrix} -\lambda & 1 & 0 & \cdots & 0 & 0 \\ 0 & -\lambda & 1 & \cdots & 0 & 0 \\ 0 & 0 & -\lambda & \cdots & 0 & 0 \\ \vdots & \vdots & \vdots & \ddots & \vdots & \vdots \\ 0 & 0 & 0 & \cdots & -\lambda & 1 \\ -p_n & -p_{n-1} & -p_{n-2} & \cdots & -p_2 & -p_1 - \lambda \end{bmatrix},$$

which implies

$$\det[A - \lambda I] = (-1)^n [\lambda^n + p_1 \lambda^{n-1} + \cdots + p_n]$$

and $\lambda^n + p_1 \lambda^{n-1} + p_2 \lambda^{n-2} + \cdots + p_n = 0$ is the well-known *characteristic* or *auxiliary equation* from elementary ODE's.

In returning to the last lemma, if A has n distinct eigenvalues $\lambda_1, \ldots, \lambda_n$ with associated eigenvectors $c_1, \ldots, c_n$, then the solutions $x_j(t) = e^{\lambda_j t} c_j$, $1 \le j \le n$ constitute a basis for the solution space of $x' = Ax$.

Question: What if $\det[A - \lambda I] = 0$ has roots of multiplicity greater than 1?

Recall that in the case of the scalar equation, if λ_0 is a root of $\lambda^n + p_1 \lambda^{n-1} + \cdots + p_n = 0$ of multiplicity k, then $e^{\lambda_0 t}, te^{\lambda_0 t}, \ldots, t^{k-1} e^{\lambda_0 t}$ are k L.I. solutions corresponding to the root λ_0, i.e., $e^{\lambda_0 t}[c_k + c_{k-1} t + \cdots + c_1 t^{k-1}]$ is a solution for all values of the scalars c_i.

In answering our question, suppose we look at the possibility of $x' = Ax$ having a solution of the form

$$x(t) = e^{\lambda_j t} \left[c_1 \frac{t^{k-1}}{(k-1)!} + c_2 \frac{t^{k-2}}{(k-2)!} + \cdots + c_{k-1} t + c_k \right],$$

where the c_i's are n-vectors and λ_j is a multiple root of the characteristic equation. To see what is involved in $x(t)$ being a solution, let us examine $x'(t) - Ax(t)$:

Now

$$x'(t) = \lambda_j e^{\lambda_j t} \left[c_1 \frac{t^{k-1}}{(k-1)!} + c_2 \frac{t^{k-2}}{(k-2)!} + \cdots + c_{k-1} t + c_k \right]$$
$$+ e^{\lambda_j t} \left[c_1 \frac{t^{k-2}}{(k-2)!} + c_2 \frac{t^{k-3}}{(k-3)!} + \cdots + c_{k-1} \right]$$

yields

$$x'(t) - Ax(t) = e^{\lambda_j(t)} \left[(\lambda_j c_1 - Ac_1) \frac{t^{k-1}}{(k-1)!} + (\lambda_j c_2 - Ac_2) \frac{t^{k-2}}{(k-2)!} \right.$$
$$\left. + \cdots + (\lambda_j c_{k-1} - Ac_{k-1}) t + (\lambda_j c_k - Ac_k) \right]$$
$$+ e^{\lambda_j t} \left[c_1 \frac{t^{k-2}}{(k-2)!} + c_2 \frac{t^{k-3}}{(k-3)!} + \cdots + c_{k-1} \right].$$

If

$$\lambda_j c_1 - Ac_1 = 0,$$
$$Ac_2 - \lambda_j c_2 = c_1,$$
$$Ac_3 - \lambda_j c_3 = c_2, \tag{5.16}$$
$$\vdots$$
$$Ac_k - \lambda_j c_k = c_{k-1},$$

then $x'(t) - Ax(t) = 0$, or $x'(t) = Ax(t)$, or $x(t)$ is a solution. We will now check to see if conditions (5.16) are indeed the case. For this, let λ be an eigenvalue of A of multiplicity $k > 1$ and write $B = A - \lambda I$. Then B is singular, because λ is an eigenvalue. Let $\mathcal{N}(B)$ be the null space of B, i.e., $\mathcal{N}(B) = \{c \in \mathbb{C}^n \,|\, Bc = 0\}$. Now $\mathcal{N}(B) \subseteq \mathcal{N}(B^2) \subseteq \mathcal{N}(B^3) \subseteq \cdots$, and since our space $\mathbb{C}^n$ is n-dimensional, $n < \infty$, thus there exists an integer $r \geq 1$ such that $\mathcal{N}(B) \subseteq \mathcal{N}(B^2) \subseteq \cdots \subseteq \mathcal{N}(B^r)$, and $\mathcal{N}(B^r) = \mathcal{N}(B^j)$, for all $j \geq r$.

Moreover, $\mathcal{N}(B^r)$ has dimension k, hence, we can find a basis $d_1, \ldots, d_k$ for $\mathcal{N}(B^r)$. Let us consider one of these basis vectors d_l. For notation, define $d_l := d_l^1$. (Let ":=" denote "defines".) Then, we define d_l^i, $i = 1, \ldots, r$, as

$$B d_l^1 = (A - \lambda I) d_l^1 = A d_l^1 - \lambda d_l^1 := d_l^2,$$
$$B^2 d_l^1 = B B d_l^1 = B d_l^2 = A d_l^2 - \lambda d_l^2 := d_l^3,$$
$$B^3 d_l^1 = B B^2 d_l^1 = B d_l^3 = A d_l^3 - \lambda d_l^3 := d_l^4, \tag{5.17}$$
$$\vdots$$
$$B^{r-1} d_l^1 = A d_l^{r-1} - \lambda d_l^{r-1} := d_l^r.$$

Now, $0 = B^r d_l^1 = A d_l^r - \lambda d_l^r$, since d_l^1 was a basis element of $\mathcal{N}(B^r)$. Comparing these results to conditions (5.16), we see that we have the same type of equations here with d_l^j playing the part of c_k (things are listed in the reverse order here from the manner listed in (5.16)).

We conclude then that

$$x(t) = e^{\lambda t}\left[d_l^r \frac{t^{r-1}}{(r-1)!} + d_l^{r-1}\frac{t^{r-2}}{(r-2)!} + \cdots + d_l^2 t + d_l^1\right] \tag{5.18}$$

is a solution of $x' = Ax$, where $d_l^1 := d_l$ is a basis vector in $\mathcal{N}(B^r)$ (note that $x(0) = d_l^1$ which must be the initial condition "this solution" satisfies). Since $\dim\mathcal{N}(B^r) = k$, there are k of these vectors d_j, each of which produces a solution corresponding to λ. Hence we get k solutions for the multiplicity of the root λ of the characteristic polynomial.

Example 5.4. Let

$$A = \left[\begin{array}{cc|cc} 0 & 1 & 0 & 0 \\ -4 & 4 & 0 & 0 \\ \hline 0 & 0 & 2 & 1 \\ 0 & 0 & 0 & 2 \end{array}\right].$$

Then,

$$A - \lambda I = \left[\begin{array}{cccc} -\lambda & 1 & 0 & 0 \\ -4 & 4-\lambda & 0 & 0 \\ 0 & 0 & 2-\lambda & 1 \\ 0 & 0 & 0 & 2-\lambda \end{array}\right],$$

which yields

$$\det[A - \lambda I] = (2-\lambda)\left|\begin{array}{ccc} -\lambda & 1 & 0 \\ -4 & 4-\lambda & 0 \\ 0 & 0 & 2-\lambda \end{array}\right| = (2-\lambda)(2-\lambda)\left|\begin{array}{cc} -\lambda & 1 \\ -4 & 4-\lambda \end{array}\right|$$

$$= (2-\lambda)^2(\lambda^2 - 4\lambda + 4) = (\lambda - 2)^4.$$

Hence, $\lambda = 2$ is an eigenvalue of the matrix A of multiplicity 4.

Let

$$B = A - \lambda I = A - 2I = \left[\begin{array}{cccc} -2 & 1 & 0 & 0 \\ -4 & 2 & 0 & 0 \\ 0 & 0 & 0 & 1 \\ 0 & 0 & 0 & 0 \end{array}\right].$$

We now determine the successive null-spaces of B, B^2, etc.

First, find the form of C such that $BC = (0)$. Well,

$$B \cdot \begin{bmatrix} c_1 \\ c_2 \\ c_3 \\ c_4 \end{bmatrix} = \begin{bmatrix} 0 \\ 0 \\ 0 \\ 0 \end{bmatrix} \iff \begin{bmatrix} -2c_1 + c_2 \\ -4c_1 + 2c_2 \\ c_4 \\ 0 \end{bmatrix} = \begin{bmatrix} 0 \\ 0 \\ 0 \\ 0 \end{bmatrix}.$$

So, $c_2 = 2c_1$, $c_4 = 0$, and c_3 is arbitrary. Hence,

$$\mathcal{N}(B) = \left\{ \begin{bmatrix} c_1 \\ 2c_1 \\ c_3 \\ 0 \end{bmatrix} \right\} \quad \text{and clearly a basis for } \mathcal{N}(B) \text{ is } \left\{ \begin{bmatrix} 1 \\ 2 \\ 0 \\ 0 \end{bmatrix}, \begin{bmatrix} 0 \\ 0 \\ 1 \\ 0 \end{bmatrix} \right\}.$$

Now

$$B^2 = \begin{bmatrix} 0 & 0 & 0 & 0 \\ 0 & 0 & 0 & 0 \\ 0 & 0 & 0 & 0 \\ 0 & 0 & 0 & 0 \end{bmatrix} \implies \mathcal{N}(B^2) = \mathbb{C}^4.$$

We could extend the basis for the $\mathcal{N}(B)$ to a basis for $\mathcal{N}(B^2) = \mathbb{C}^4$ or we could find another basis for $\mathbb{C}^4$.

Take $\{e_1, e_2, e_3, e_4\}$ as our basis for $\mathcal{N}(B^2)$. Apply B to each of these basis vectors as is done in the calculation above denoted by (5.17). Since B^2 (any vector) $= (0)$, from the form of a solution given in (5.18), four solutions will be of the form (note: $r = 2$, $\lambda = 2$ relative to (5.17))

$$x_j(t) = e^{\lambda t}[Be_j t + e_j].$$

We have

$$Be_1 = \begin{bmatrix} -2 \\ -4 \\ 0 \\ 0 \end{bmatrix}, \quad Be_2 = \begin{bmatrix} 1 \\ 2 \\ 0 \\ 0 \end{bmatrix}, \quad Be_3 = \begin{bmatrix} 0 \\ 0 \\ 0 \\ 0 \end{bmatrix}, \quad Be_4 = \begin{bmatrix} 0 \\ 0 \\ 1 \\ 0 \end{bmatrix},$$

with corresponding solutions

$$x_1(t) = e^{\lambda t} \left(Be_1 t + e_1 \right) = e^{2t} \begin{bmatrix} -2t+1 \\ -4t \\ 0 \\ 0 \end{bmatrix},$$

$$x_2(t) = e^{\lambda t} \left(Be_2 t + e_2 \right) = e^{2t} \begin{bmatrix} t \\ 2t+1 \\ 0 \\ 0 \end{bmatrix},$$

$$x_3(t) = e^{\lambda t} \left(Be_3 t + e_3 \right) = e^{2t} \begin{bmatrix} 0 \\ 0 \\ 1 \\ 0 \end{bmatrix},$$

$$x_4(t) = e^{\lambda t} \left(Be_4 t + e_4 \right) = e^{2t} \begin{bmatrix} 0 \\ 0 \\ t \\ 1 \end{bmatrix}.$$

As noted before,

$$x_j(0) = e_j, \ 1 \leq j \leq 4 \ \text{(always a basis element of } \mathcal{N}(B^r)\text{)}$$

$$= d_l^1 \ \text{(as denoted in our formula in (5.17))}$$

From this, we have the fundamental matrix of the system

$$\begin{cases} X' = AX, \\ X(0) = I, \end{cases}$$

as

$$e^{tA} = e^{2t} \begin{bmatrix} -2t+1 & t & 0 & 0 \\ -4t & 2t+1 & 0 & 0 \\ 0 & 0 & 1 & t \\ 0 & 0 & 0 & 1 \end{bmatrix}, \quad \text{(since } e^{0 \cdot A} = I \text{ in this case)}.$$

Now, there is an alternate method for finding 4 L.I. solutions of $x' = Ax$, with A as above. Consider the basis spanning

$$\mathcal{N}(B) = \left\{ \begin{bmatrix} 1 \\ 2 \\ 0 \\ 0 \end{bmatrix} \begin{bmatrix} 0 \\ 0 \\ 1 \\ 0 \end{bmatrix} \right\}.$$

We can extend this to a basis of $\mathcal{N}(B^2) = \mathbb{C}^4$ by

$$\mathcal{N}(B^2) = \mathbb{C}^4 = \left\{ \begin{bmatrix} 1 \\ 2 \\ 0 \\ 0 \end{bmatrix}, \begin{bmatrix} 0 \\ 0 \\ 1 \\ 0 \end{bmatrix}, \begin{bmatrix} 0 \\ 0 \\ 0 \\ 1 \end{bmatrix}, \begin{bmatrix} 1 \\ 0 \\ 0 \\ 0 \end{bmatrix} \right\}.$$

Corresponding to the first two basis elements, since

$$B \begin{bmatrix} 1 \\ 2 \\ 0 \\ 0 \end{bmatrix} = \begin{bmatrix} 0 \\ 0 \\ 0 \\ 0 \end{bmatrix} \quad \text{and} \quad B \begin{bmatrix} 0 \\ 0 \\ 1 \\ 0 \end{bmatrix} = \begin{bmatrix} 0 \\ 0 \\ 0 \\ 0 \end{bmatrix},$$

we have the two solutions

$$x_1(t) = e^{2t} \begin{bmatrix} 1 \\ 2 \\ 0 \\ 0 \end{bmatrix}, \quad x_2(t) = e^{2t} \begin{bmatrix} 0 \\ 0 \\ 1 \\ 0 \end{bmatrix}.$$

We can also obtain $x_3(t)$ and $x_4(t)$ by applying B to

$$\begin{bmatrix} 0 \\ 0 \\ 0 \\ 1 \end{bmatrix} \quad \text{and} \quad \begin{bmatrix} 1 \\ 0 \\ 0 \\ 0 \end{bmatrix},$$

so that

$$x_3(t) = e^{2t} \left(B \begin{bmatrix} 0 \\ 0 \\ 0 \\ 1 \end{bmatrix} t + \begin{bmatrix} 0 \\ 0 \\ 0 \\ 1 \end{bmatrix} \right), \quad x_4(t) = e^{2t} \left(B \begin{bmatrix} 1 \\ 0 \\ 0 \\ 0 \end{bmatrix} t + \begin{bmatrix} 1 \\ 0 \\ 0 \\ 0 \end{bmatrix} \right).$$

Here, our $x_1(t)$ and $x_2(t)$ are linear combinations of those found using the first method.

$\boxed{\text{Exercise}}$ **34.** Given A, find 4 linearly independent solutions of $x' = Ax$ for each of the following:

(1) $\begin{bmatrix} -1 & -1 & -1 & 2 \\ 0 & 1 & 0 & 0 \\ -1 & -1 & 0 & 1 \\ -1 & 0 & -1 & 2 \end{bmatrix}$;   (2) $\begin{bmatrix} 1 & -1 & -1 & 1 \\ 1 & 2 & 1 & -1 \\ 0 & 0 & 1 & 0 \\ 1 & 0 & 0 & 1 \end{bmatrix}$;

(3) $\begin{bmatrix} 0 & 1 & 0 & -1 \\ 0 & 0 & 0 & 1 \\ 1 & -1 & 0 & 1 \\ 0 & 0 & 0 & 1 \end{bmatrix}$; (4) $\begin{bmatrix} 1 & 0 & 0 & 1 \\ 1 & 3 & 1 & -1 \\ -1 & -1 & 2 & 1 \\ 0 & 1 & 1 & 2 \end{bmatrix}$.

Definition 5.5. $A, B \in \mathcal{M}_n$ are said to be *similar* if there exists a non-singular matrix C such that $C^{-1}AC = B$. (This defines an equivalence relation on $\mathcal{M}_n$.)

Theorem 5.7. *Every matrix* $A \in \mathcal{M}_n$ *is similar to a block diagonal matrix*

$$J = \begin{bmatrix} J_0 & & & 0 \\ & J_1 & & \\ & & \ddots & \\ 0 & & & J_s \end{bmatrix},$$

where J_0 *is the diagonal matrix*

$$J_0 = \begin{bmatrix} \lambda_1 & 0 & 0 & \cdots & 0 \\ 0 & \lambda_2 & 0 & \cdots & 0 \\ \vdots & \vdots & \vdots & \ddots & \vdots \\ 0 & 0 & 0 & \cdots & \lambda_q \end{bmatrix},$$

and each of the blocks J_i, $1 \le i \le s$, *is of the form*

$$J_i = \begin{bmatrix} \lambda_{q+i} & 1 & 0 & \cdots & 0 & 0 \\ 0 & \lambda_{q+i} & 1 & \cdots & 0 & 0 \\ 0 & 0 & \lambda_{q+i} & \ddots & 0 & 0 \\ \vdots & \vdots & \vdots & \ddots & \ddots & \vdots \\ 0 & 0 & 0 & \cdots & \lambda_{q+i} & 1 \\ 0 & 0 & 0 & \cdots & 0 & \lambda_{q+i} \end{bmatrix},$$

where λ_{q+i}*'s on the diagonal are the same and* 1*'s are on the super diagonal.*

The number $\lambda_1, \lambda_2, \ldots, \lambda_{q+s}$ are the eigenvalues of A and are repeated in J according to their multiplicities. Furthermore, if λ_j is a simple eigenvalue, then λ_j appears in J_0, hence, if all the eigenvalues are simple, J is a diagonal matrix. The matrix J is called the *"Jordan Canonical Form* of A" and is unique up to rearrangements of the blocks J_i, $0 \le i \le s$, along the diagonal. (So, there exists a C such that $C^{-1}AC = J$.)

Let us look at the previous example and see how the Jordan form relates. Recall $x' = Ax$, where

$$A = \begin{bmatrix} 0 & 1 & 0 & 0 \\ -4 & 4 & 0 & 0 \\ 0 & 0 & 2 & 1 \\ 0 & 0 & 0 & 2 \end{bmatrix}.$$

We found that $\det[A - \lambda I] = (\lambda - 2)^4$ and

$$B = A - 2I = \begin{bmatrix} -2 & 1 & 0 & 0 \\ -4 & 2 & 0 & 0 \\ 0 & 0 & 0 & 1 \\ 0 & 0 & 0 & 0 \end{bmatrix}, \quad \mathcal{N}(B) = \left\{ \begin{bmatrix} 1 \\ 2 \\ 0 \\ 0 \end{bmatrix}, \begin{bmatrix} 0 \\ 0 \\ 1 \\ 0 \end{bmatrix} \right\},$$

and $B^2 = (0)$. Extending the basis for $\mathcal{N}(B)$ to a basis for $\mathcal{N}(B^2) = \mathbb{C}^4$, we obtained

$$\mathcal{N}(B^2) = \left\{ \underset{c_1}{\begin{bmatrix} 1 \\ 2 \\ 0 \\ 0 \end{bmatrix}}, \underset{d_1}{\begin{bmatrix} 0 \\ 0 \\ 1 \\ 0 \end{bmatrix}}, \underset{c_2}{\begin{bmatrix} 1 \\ 0 \\ 0 \\ 0 \end{bmatrix}}, \underset{d_2}{\begin{bmatrix} 0 \\ 0 \\ 0 \\ 1 \end{bmatrix}} \right\}.$$

Applying B to c_2 and d_2,

$$B \begin{bmatrix} 1 \\ 0 \\ 0 \\ 0 \end{bmatrix} = \begin{bmatrix} -2 \\ -4 \\ 0 \\ 0 \end{bmatrix}, \quad B \begin{bmatrix} 0 \\ 0 \\ 0 \\ 1 \end{bmatrix} = \begin{bmatrix} 0 \\ 0 \\ 1 \\ 0 \end{bmatrix},$$

we have $Bc_2 = -2c_1$ and $Bd_2 = d_1$. Hence $[A - 2I]c_2 = -2c_1$ and or, $Ac_2 = 2c_2 - 2c_1$. On the other hand, $Bc_1 = [0] \iff [A - 2I]c_1 = 0$. So, $Ac_1 = 2c_1$.

Now $Bd_2 = d_1$ yields $Ad_2 = 2d_2 + d_1$ and $Bd_1 = [0]$ yields $Ad_1 = 2d_1$. Form C by taking c_1, c_2, d_1, d_2 as its columns; i.e., $C = [c_1, c_2, d_1, d_2]_{4\times4}$.

Now

$$\begin{aligned} C^{-1}AC &= C^{-1}[Ac_1, Ac_2, Ad_1, Ad_2] \\ &= C^{-1}[2c_1, 2c_2 - 2c_1, 2d_1, 2d_2 + d_1] \\ &= [2C^{-1}c_1, 2C^{-1}c_2 - 2C^{-1}c_1, 2C^{-1}d_1, 2C^{-1}d_2 + C^{-1}d_1] \\ &= [2e_1, 2e_2 - 2e_1, 2e_3, 2e_4 + e_3] \\ &= \begin{bmatrix} 2 & -2 & 0 & 0 \\ 0 & 2 & 0 & 0 \\ 0 & 0 & 2 & 1 \\ 0 & 0 & 0 & 2 \end{bmatrix}, \end{aligned}$$

where $C^{-1}C = I$ because

$$C^{-1}C = C^{-1}[c_1, c_2, d_1, d_2]$$
$$= [C^{-1}c_1, C^{-1}c_2, C^{-1}d_1, C^{-1}d_2]$$
$$= I = [e_1, e_2, e_3, e_4].$$

This is almost the Jordan form. We need to normalize the -2 entry. Thus, we return to the point where we extended our basis and we choose

$$c_2 = \begin{bmatrix} -\frac{1}{2} \\ 0 \\ 0 \\ 0 \end{bmatrix}.$$

Then, $[A - 2I]c_2 = c_1$. So, $Ac_2 = 2c_2 + c_1$ and continuing as above, we obtain

$$
\begin{aligned}
C^{-1}AC &= C^{-1}[2c_1, 2c_2 + c_1, 2d_1, 2d_2 + d_1] \\
&= [2C^{-1}c_1, 2C^{-1}c_2 + C^{-1}c_1, 2C^{-1}d_1, 2C^{-1}d_2 + C^{-1}d_1] \\
&= [2e_1, 2e_2 + e_1, 2e_3, 2e_4 + e_3] \\
&= \begin{bmatrix} 2 & 1 & 0 & 0 \\ 0 & 2 & 0 & 0 \\ 0 & 0 & 2 & 1 \\ 0 & 0 & 0 & 2 \end{bmatrix}.
\end{aligned}
$$

This is the Jordan canonical form of A.

Consider now the matrix IVP: $X' = AX$, $X(0) = I$, where A is a constant matrix. We have previously shown that the solution is given by $X(t) = e^{tA}$. Let C be a nonsingular matrix such that $C^{-1}AC = J$, the Jordan canonical form of A.

Let $Y(t) = C^{-1}X(t)C = C^{-1}e^{tA}C$. Then

$$
\begin{aligned}
Y'(t) &= C^{-1}X'(t)C = C^{-1}AX(t)C \\
&= C^{-1}ACC^{-1}X(t)C \\
&= J\,Y(t).
\end{aligned}
$$

Furthermore, $Y(0) = C^{-1}IC = I$. So, $Y(t)$ is the solution of the IVP

$$\begin{cases} Y' = JY, \\ Y(0) = I. \end{cases}$$

Hence, $Y(t) = e^{tJ}$ and

$$C^{-1}e^{tA}C = Y(t) = e^{tJ} = e^{t\,C^{-1}AC}, \quad (\text{or } Ce^{tJ}C^{-1} = e^{tCJC^{-1}} = e^{tA}).$$

Hence, if $X(t) = CY(t)C^{-1}$, and if we can find $Y(t) = e^{tJ}$, we have an idea of what $X(t) = e^{tA}$ looks like.

Let calculate e^{tJ}. First

$$e^{tJ} = I + tJ + \frac{(tJ)^2}{2!} + \frac{(tJ)^3}{3!} + \cdots,$$

where

$$J = \begin{bmatrix} J_0 & & & 0 \\ & J_1 & & \\ & & \ddots & \\ 0 & & & J_s \end{bmatrix} \implies J^j = \begin{bmatrix} J_0^j & & & 0 \\ & J_1^j & & \\ & & \ddots & \\ 0 & & & J_s^j \end{bmatrix}.$$

Thus

$$e^{tJ} = I + \begin{bmatrix} tJ_0 & & & 0 \\ & tJ_1 & & \\ & & \ddots & \\ 0 & & & tJ_s \end{bmatrix} + \begin{bmatrix} \frac{t^2 J_0^2}{2!} & & & 0 \\ & \frac{t^2 J_1^2}{2!} & & \\ & & \ddots & \\ 0 & & & \frac{t^2 J_s^2}{2!} \end{bmatrix} + \cdots$$

$$= \begin{bmatrix} e^{tJ_0} & & & 0 \\ & e^{tJ_1} & & \\ & & \ddots & \\ 0 & & & e^{tJ_s} \end{bmatrix}.$$

Hence, calculating e^{tJ} is reduced to calculating e^{tJ_i} for each block.

First,

$$J_0 = \begin{bmatrix} \lambda_1 & & & 0 \\ & \lambda_2 & & \\ & & \ddots & \\ 0 & & & \lambda_q \end{bmatrix} \implies J_0^j = \begin{bmatrix} \lambda_1^j & & & 0 \\ & \lambda_2^j & & \\ & & \ddots & \\ 0 & & & \lambda_q^j \end{bmatrix}.$$

After summing, we get

$$e^{tJ_0} = \begin{bmatrix} e^{\lambda_1 t} & & & 0 \\ & e^{\lambda_2 t} & & \\ & & \ddots & \\ 0 & & & e^{\lambda_q t} \end{bmatrix}.$$

Next, consider an arbitrary block,

$$
J_i = \begin{bmatrix}
\lambda_{q+i} & 1 & 0 & \cdots & 0 & 0 \\
0 & \lambda_{q+i} & 1 & \cdots & 0 & 0 \\
0 & 0 & \lambda_{q+i} & \cdots & 0 & 0 \\
\vdots & \vdots & \vdots & \ddots & \vdots & \vdots \\
0 & 0 & 0 & \cdots & \lambda_{q+i} & 1 \\
0 & 0 & 0 & \cdots & 0 & \lambda_{q+i}
\end{bmatrix}.
$$

For brevity, let $\lambda = \lambda_{q+i}$, so that

$$
J_i = \begin{bmatrix}
\lambda & 0 & 0 & \cdots & 0 \\
0 & \lambda & 0 & \cdots & 0 \\
0 & 0 & \lambda & \cdots & 0 \\
\vdots & \vdots & \vdots & \ddots & \vdots \\
0 & 0 & 0 & \cdots & \lambda
\end{bmatrix} + \begin{bmatrix}
0 & 1 & 0 & \cdots & 0 \\
0 & 0 & 1 & \cdots & 0 \\
\vdots & \vdots & \vdots & \ddots & \vdots \\
0 & 0 & 0 & \cdots & 1 \\
0 & 0 & 0 & \cdots & 0
\end{bmatrix} := \lambda I_m + Z_m.
$$

Now

$$
Z_m^2 = \begin{bmatrix}
0 & 0 & 1 & 0 & \cdots & 0 \\
0 & 0 & 0 & 1 & \cdots & 0 \\
\vdots & \vdots & \vdots & \vdots & \ddots & \vdots \\
0 & 0 & 0 & 0 & \cdots & 1 \\
0 & 0 & 0 & 0 & \cdots & 0 \\
0 & 0 & 0 & 0 & \cdots & 0
\end{bmatrix} \quad \text{and} \quad Z_m^{m-1} = \begin{bmatrix}
0 & 0 & \cdots & 1 \\
0 & 0 & \cdots & 0 \\
\vdots & \vdots & \ddots & \vdots \\
0 & 0 & \cdots & 0
\end{bmatrix}
$$

yield $Z_m^m = [0]$. Hence

$$
\begin{aligned}
e^{tJ_i} &= e^{t\lambda I_m + tZ_m} = e^{t\lambda I_m} e^{tZ_m} \quad \text{(since } I_m \text{ commutes with any matrix)} \\
&= e^{t\lambda} e^{tZ_m} \quad \text{(because } I_m Z_m = Z_m) \\
&= e^{\lambda t} \sum_{k=0}^{m-1} \frac{t^k Z_m^k}{k!} \\
&= e^{\lambda t} \begin{bmatrix}
1 & t & \frac{t^2}{2!} & \cdots & \frac{t^{m-2}}{(m-2)!} & \frac{t^{m-1}}{(m-1)!} \\
0 & 1 & t & \cdots & \frac{t^{m-3}}{(m-3)!} & \frac{t^{m-2}}{(m-2)!} \\
0 & 0 & 1 & \cdots & \frac{t^{m-4}}{(m-4)!} & \frac{t^{m-3}}{(m-3)!} \\
\vdots & \vdots & \vdots & \ddots & \vdots & \vdots \\
0 & 0 & 0 & \cdots & 1 & t \\
0 & 0 & 0 & \cdots & 0 & 1
\end{bmatrix},
\end{aligned}
$$

that is,

$$e^{tJ_j} = \begin{bmatrix} e^{t\lambda_{q+i}} & te^{t\lambda_{q+i}} & \frac{t^2}{2!}e^{t\lambda_{q+i}} & \cdots & \frac{t^{m-1}}{(m-1)!}e^{t\lambda_{q+i}} \\ 0 & e^{t\lambda_{q+i}} & te^{t\lambda_{q+i}} & \cdots & \frac{t^{m-2}}{(m-2)!}e^{t\lambda_{q+i}} \\ 0 & 0 & e^{t\lambda_{q+i}} & \cdots & \frac{t^{m-3}}{(m-3)!}e^{t\lambda_{q+i}} \\ \vdots & \vdots & \vdots & \vdots & \vdots \\ 0 & 0 & 0 & \cdots & te^{t\lambda_{q+i}} \\ 0 & 0 & 0 & \cdots & e^{t\lambda_{q+i}} \end{bmatrix}.$$

Thus, we know what each block of e^{tJ} looks like. Since $e^{tA} = Ce^{tJ}C^{-1}$ where C is a nonsingular constant matrix, the elements in e^{tA} are of the form $\sum_{j=1}^{q+s} p_j(t)e^{\lambda_j t}$, where each $p_j(t)$ is a polynomial in t. Furthermore, if the largest of the blocks, J_i, $1 \le i \le s$, (not J_0), is an $m \times m$ matrix, then all polynomials appearing in the elements of e^{tA} are of degree $\le m - 1$.

Since each solution of $x' = Ax$ is of the form $x(t) = e^{tA}c_0$, where $c_0 \in \mathbb{C}^n$, the elements of the vector $x(t)$ are also of the form $\sum_{j=1}^{q+s} p_j(t)e^{\lambda_j t}$. This leads to the following remarks concerning solutions of $x' = Ax$.

Remark 5.5.

(1) If $\mathrm{Re}\,\lambda_j < 0$, for each eigenvalue λ_j of A, then all solutions $x(t)$ are such that $\|x(t)\|$ is bounded on $[0, +\infty)$ and $\|x(t)\| \to 0$ as $t \to +\infty$. Reason: Using the max norm, where $x(t)$ is a vector, if we look at a piece of the polynomial of some entry, $|t^r e^{\lambda_j t}| \le t^r e^{\mathrm{Re}\lambda_j t} \to 0$, if $\mathrm{Re}\lambda_j < 0$, as $t \to +\infty$. So, as $t \to +\infty$,

$$\max \left| \sum p_j(t)e^{\lambda_j t} \right| \to 0.$$

(2) If $\mathrm{Re}\,\lambda_j \le 0$, for each eigenvalue λ_j of A, then all the solutions $x(t)$ are bounded on $[0, +\infty)$ iff the λ_j's for which $\mathrm{Re}\lambda_j = 0$ appear only in J_0, the diagonal block in the Jordan canonical form of A. Reason: $e^{\lambda_j t} = e^{iwt} = \cos wt + i \sin wt$, for $\mathrm{Re}\lambda_j = 0$. Thus the only way for $\|x(t)\|$ to be unbounded is that an element other than zero must be off the diagonal (where $\mathrm{Re}\lambda_j \le 0$).

(3) If $\mathrm{Re}\,\lambda_j > 0$, for some eigenvalues of A, then $x' = Ax$ has some solutions which are unbounded on $[0, +\infty)$.

5.7   The Logarithm of a Matrix

Let's now consider another question. If $A \in \mathcal{M}_n$, e^{tA} is a nonsingular matrix, for all $t \in \mathbb{R}$, and in particular e^A is nonsingular.

Question: If A is nonsingular, is there a $B \in \mathcal{M}_n$ such that $e^B = A$? If so, we will call B a *logarithm* of A; i.e., $B = \log A$.

Let J be the Jordan canonical form of A and assume $C^{-1}AC = J$. Then if $e^D = J$, it follows that $Ce^D C^{-1} = CJC^{-1} = A$. So, $e^{CDC^{-1}} = A$. If $D = \log J$, then $CDC^{-1} = \log A$. Furthermore, if D is in block form, say

$$D = \begin{bmatrix} D_0 & & & 0 \\ & D_1 & & \\ & & \ddots & \\ 0 & & & D_s \end{bmatrix},$$

then

$$e^D = \begin{bmatrix} e^{D_0} & & & 0 \\ & e^{D_1} & & \\ & & \ddots & \\ 0 & & & e^{D_s} \end{bmatrix}.$$

Now we want $e^D = J$, thus it suffices to determine $D_0, D_1, \ldots, D_s$ such that $e^{D_i} = J_i$, $0 \le i \le s$.

Note that, if A is nonsingular, then all the eigenvalues are nonzero, and hence, each eigenvalue has a logarithm (in the complex sense, $\log \lambda = \ln |\lambda| + i \arg \lambda$).

Now

$$J_0 = \begin{bmatrix} \lambda_1 & & & 0 \\ & \lambda_2 & & \\ & & \ddots & \\ 0 & & & \lambda_q \end{bmatrix} \implies D_0 = \begin{bmatrix} \log \lambda_1 & & & 0 \\ & \log \lambda_2 & & \\ & & \ddots & \\ 0 & & & \log \lambda_q \end{bmatrix},$$

because

$$e^{D_0} = \begin{bmatrix} e^{\log \lambda_1} & & & 0 \\ & e^{\log \lambda_2} & & \\ & & \ddots & \\ 0 & & & e^{\log \lambda_q} \end{bmatrix} = J_0.$$

Now we look at a typical block,

$$J_i = \begin{bmatrix} \lambda & 1 & 0 & \cdots & 0 & 0 \\ 0 & \lambda & 1 & \cdots & 0 & 0 \\ 0 & 0 & \lambda & \cdots & 0 & 0 \\ \vdots & \vdots & \vdots & \ddots & \vdots & \vdots \\ 0 & 0 & 0 & \cdots & \lambda & 1 \\ 0 & 0 & 0 & \cdots & 0 & \lambda \end{bmatrix} = \lambda I_m + Z_m = (\lambda I_m)\left(I_m + \frac{1}{\lambda} Z_m\right).$$

Now λI_m is diagonal, so by techniques used with D_0 and J_0, $\log(\lambda I_m) = (\log \lambda)I_m$. If we can determine

$$\log\left(I_m + \frac{1}{\lambda}I_m\right), \tag{5.19}$$

then it and $(\log \lambda)I_m$ commute and we will have

$$e^{\log(\lambda I_m) + \log\left(I_m + \frac{1}{\lambda}Z_m\right)} = e^{(\log \lambda I_m)}e^{\log\left(I_m + \frac{1}{\lambda}Z_m\right)}$$

$$= (\lambda I_m)\left(I_m + \frac{1}{\lambda}Z_m\right) = J_i.$$

Thus the problem of determining $\log J_i$ is reduced to determining $\log\left(I_m + \frac{1}{\lambda}Z_m\right)$. Recalling that $\log(1 + x) = \sum_{k=1}^{\infty} \frac{(-1)^{k+1}x^k}{k}$ is absolutely convergent for $|x| < 1$ and recalling the form of Z_m, we have **formally** that

$$\log\left(I_m + \frac{1}{\lambda}Z_m\right) = \sum_{k=1}^{m-1} \frac{(-1)^{k+1}}{k}\left(\frac{Z_m}{\lambda}\right)^k. \tag{5.20}$$

This is a finite sum, since $Z_m^m = [0]$. Thus,

$$e^{\sum_{k=1}^{m-1} \frac{(-1)^{k+1}}{k}\left(\frac{Z_m}{\lambda}\right)^k} = I_m + \frac{1}{\lambda}Z_m.$$

Thus, it is the case that (5.20) is obtained. Consequently,

$$\log J_i = (\log \lambda)I_m + \sum_{k-1}^{m-1} \frac{(-1)^{k+1}}{k}\left(\frac{Z_m}{\lambda}\right)^k.$$

In ending our discussion of constant coefficient matrix theory, let's apply this last concept to our previous example $x' = Ax$, where

$$A = \begin{bmatrix} 0 & 1 & 0 & 0 \\ -4 & 4 & 0 & 0 \\ 0 & 0 & 2 & 1 \\ 0 & 0 & 0 & 2 \end{bmatrix}.$$

We calculated the matrix C which transforms A into Jordan form to be

$$C = \begin{bmatrix} 1 & -\frac{1}{2} & 0 & 0 \\ 2 & 0 & 0 & 0 \\ 0 & 0 & 1 & 0 \\ 0 & 0 & 0 & 1 \end{bmatrix},$$

and we obtained

$$C^{-1} = \begin{bmatrix} 0 & \frac{1}{2} & 0 & 0 \\ -2 & 1 & 0 & 0 \\ 0 & 0 & 1 & 0 \\ 0 & 0 & 0 & 1 \end{bmatrix}, \quad C^{-1}AC = \begin{bmatrix} 2 & 1 & 0 & 0 \\ 0 & 2 & 0 & 0 \\ 0 & 0 & 2 & 1 \\ 0 & 0 & 0 & 2 \end{bmatrix},$$

since the blocks of the Jordon form are the same, we need only to calculate the logarithm for one block.

Note: $m = 2$ and $\lambda = 2$, so

$$Z_m = Z_2 = \begin{bmatrix} 0 & 1 \\ 0 & 0 \end{bmatrix}, \quad \frac{Z_m}{\lambda} = \begin{bmatrix} 0 & \frac{1}{2} \\ 0 & 0 \end{bmatrix}.$$

Hence,

$$\log \begin{bmatrix} 2 & 1 \\ 0 & 2 \end{bmatrix} = (\ln 2) \begin{bmatrix} 1 & 0 \\ 0 & 1 \end{bmatrix} + \frac{(-1)^2}{1} \begin{bmatrix} 0 & \frac{1}{2} \\ 0 & 0 \end{bmatrix} = \begin{bmatrix} \ln 2 & \frac{1}{2} \\ 0 & \ln 2 \end{bmatrix}.$$

Therefore,

$$C \begin{bmatrix} \ln 2 & \frac{1}{2} & 0 & 0 \\ 0 & \ln 2 & 0 & 0 \\ 0 & 0 & \ln 2 & \frac{1}{2} \\ 0 & 0 & 0 & \ln 2 \end{bmatrix} C^{-1} = \log A.$$

Chapter 6

Periodic Linear Systems and Floquet Theory

6.1 Periodic Homogeneous Linear Systems and Floquet Theory

Definition 6.1. A matrix-valued function $A : (-\infty, +\infty) \to \mathcal{M}_n$ is said to have *period* $\omega > 0$ if $A(t + \omega) = A(t)$, for all $t \in \mathbb{R}$.

Theorem 6.1 (Floquet). *Let a matrix-valued function $A : (-\infty, \infty) \to \mathcal{M}_n$ be continuous and have period $\omega > 0$. Then a fundamental matrix solution for the linear system $X = A(t)X$ can be written in the form $X(t) = Y(t)e^{tR}$, where $R \in \mathcal{M}_n$ is a constant matrix and $Y(t) : (-\infty, +\infty) \to \mathcal{M}_n$ is continuous and has period ω. Furthermore, $Y(t)$ is nonsingular on $(-\infty, \infty)$.*

Proof. Let $X(t)$ be a fundamental matrix solution of $X' = A(t)X$. Then by the chain rule

$$\frac{d}{dt}X(t + \omega) = X'(t + \omega) \equiv A(t + \omega)X(t + \omega) = A(t)X(t + \omega).$$

Hence, $X(t + w)$ is also a solution of $X' = A(t)X$. By the previous theory, there exists a constant $C \in \mathcal{M}_n$ such that $X(t + \omega) = X(t)C$; i.e., these solutions are constant multiples of each other. We can determine C as follows: $X(0 + \omega) = X(0)C$ yields $C = X^{-1}(0)X(\omega)$. C is nonsingular and consequently C has a logarithm. Hence, there exists a constant matrix R such that $C = e^{\omega R}$ (actually, $R = \frac{1}{\omega}\log C$).

Define

$$Y(t) := X(t)e^{-tR}.$$

Obviously, $Y(t)$ has the following properties:

(1) $Y(t)$ is continuous on $\mathbb{R}$.

(2) $Y(t)$ is nonsingular.

(3) $Y(t)$ is periodic of periodic ω, i.e.,

$$
\begin{aligned}
Y(t+\omega) &= X(t+\omega)e^{-(t+\omega)R} \\
&= X(t)Ce^{-\omega R}e^{-tR} \\
&= X(t)CC^{-1}e^{-tR} \\
&= X(t)e^{-tR} = Y(t). \qquad \square
\end{aligned}
$$

Remark 6.1. Since $Y(t)$ is continuous on $\mathbb{R}$ and periodic, each entry in $Y(t)$ is continuous and periodic, hence bounded. So, $\|Y(t)\|$ is bounded on $\mathbb{R}$. Moreover, by the way the determinant is defined, $\det Y(t)$ is periodic. Since $\det Y(t) \neq 0$, $\det Y(t)$ is bounded away from zero, and since it is periodic, then it is bounded. Consequently $\|Y^{-1}(t)\|$ is bounded.

Definition 6.2. Let $A(t)$ and $B(t)$ be defined on $[a, +\infty)$. Then $A(t)$ is said to be *kinematically similar to* $B(t)$ on $[a, +\infty)$ in case there is an absolutely continuous (we will think differentiable) nonsingular matrix function $L(t)$ on $[a, +\infty)$ such that $\|L(t)\|$ and $\|L^{-1}(t)\|$ are both bounded on $[a, +\infty)$ and such that the transformation $x = L(t)y$ transforms the system $x' = A(t)x$ into the system $y' = B(t)y$; i.e., $L^{-1}(t)A(t)L(t) - L^{-1}(t)L'(t) \equiv B(t)$ on $[a, +\infty)$.

$\boxed{\text{Exercise}}$ **35.** (i) Show that kinematic similarity is an equivalence relation on the set of all continuous $n \times n$ matrix functions on $[a, +\infty)$.

(ii) Show that if $A(t)$ is continuous and has period ω on $(-\infty, \infty)$, then $A(t)$ is kinematically similar to a constant matrix on $\mathbb{R}$.

Note: Kinematic similarity can be thought of as an extension of similarity, by taking L to be a constant matrix.

Now let $A(t)$ be continuous and ω-periodic on $\mathbb{R}$. If $X(t)$ is a fundamental matrix solution of $X' = A(t)X$, then there is a nonsingular matrix C such that $X(t+\omega) = X(t)C$, for all $t \in \mathbb{R}$.

Remark 6.2. The matrix C is not uniquely determined by $A(t)$, however C is unique up to similarity.

Proof. Let $X(t)$ and $\Phi(t)$ both be fundamental matrix solutions of $X' = A(t)X$. Then there are matrices C and D such that $X(t+\omega) = X(t)C$ and

$\Phi(t + \omega) = \Phi(t)D$. Furthermore, there is a nonsingular matrix B such that $X(t) = \Phi(t)B$, since they are both fundamental matrix solutions. So,

$$\begin{aligned} X(t)C &= X(t + \omega) = \Phi(t + \omega)B = \Phi(t)DB \\ &= \Phi(t)BB^{-1}DB \\ &= X(t)B^{-1}DB. \end{aligned}$$

Hence, $C = B^{-1}DB$; i.e., C and D are similar. □

It follows then that $A(t)$ *does* uniquely determine the eigenvalues of C, because similar matrices have the same eigenvalues. In fact, these eigenvalues are called the *characteristic multipliers* of the periodic system $X' = A(t)X$.

For the above matrix C, where $X(t + \omega) = X(t)C$, if $\sigma_1, \ldots, \sigma_n$ are the eigenvalues of C and $\lambda_1, \ldots, \lambda_n$ are the eigenvalues of R, where $C = e^{\omega R}$ (i.e., $R = \frac{1}{\omega}\log C$), then if both sets of eigenvalues are properly ordered, we will have that $\sigma_j = e^{\omega \lambda_j}$, $1 \leq j \leq n$. The numbers λ_j are not uniquely determined, but are determined up to additive integer multiples of $\frac{2\pi i}{\omega}$. The numbers λ_j, $1 \leq j \leq n$, are called the *characteristic exponents* of the periodic system $X' = A(t)X$.

Remark 6.3. Corresponding to each eigenvalue σ of C, where $X(t + \omega) = X(t)C$, there exists a nontrivial vector solution $x(t)$ of the vector system $x' = A(t)x$ such that $x(t + w) = \sigma x(t)$ for all $t \in \mathbb{R}$.

Proof. Let σ be an eigenvalue of C and let x_0 be an associated eigenvector, then $Cx_0 = \sigma x_0$. Let $x(t)$ be a solution of $x' = A(t)x$ given by $x(t) \equiv X(t)x_0$. Then,

$$x(t + \omega) = X(t + \omega)x_0 = X(t)Cx_0 = X(t)\sigma x_0 = \sigma X(t)x_0 = \sigma x(t). \quad \square$$

Remark 6.4. This last remark tells us when we can obtain periodic solutions of a periodic system; i.e., if 1 is an eigenvalue of C, then the system $x' = A(t)x$ has a nontrivial ω-periodic solution.

Suppose σ is an nth root of unity and σ is an eigenvalue of C. Then there exists a nontrivial solution $x(t)$ with $x(t + \omega) = \sigma x(t)$. We iterate this; that is, replace "t" by "$t + \omega$". Hence, $x(t + 2\omega) = \sigma x(t + \omega) = \sigma^2 x(t), \ldots, x(t + n\omega) = \sigma^n x(t) = x(t)$, (since $\sigma^n = 1$). Thus, in the case, $x' = A(t)x$ has an $n\omega$-periodic solution.

Remark 6.5. For our next observation, if σ is an eigenvalue of C and $|\sigma| \leq 1$, then $x' = A(t)x$ has a nontrivial solution $x(t)$ which is bounded on $[0, \infty)$. (Note: C is nonsingular, thus $\sigma \neq 0$ and σ^{-1} exists.) To see the bounded part, let $t \in \mathbb{R}$ be arbitrary, but fixed, and let $\omega \in \mathbb{R}$. See the Figure 6.1.

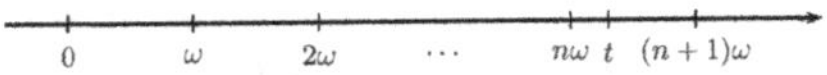

Fig. 6.1   The points $k\omega$, $k = 0, 1, 2, \ldots$.

Then, there exists $n \in \mathbb{Z}$ such that $n\omega \leq t \leq (n+1)\omega$. Therefore, $t - n\omega \in [0, \omega]$.

Using our above iteration technique, $x(t) = x((t - n\omega) + n\omega) = \sigma^n x(t - n\omega)$. So, $\|x(t)\| = |\sigma|^n \|x(t - n\omega)\| \leq M$, where $M = \max_{0 \leq t \leq \omega} \|x(t)\|$.

Note: If $|\sigma| = 1$, then there is a solution bounded on all of $\mathbb{R}$.

Example 6.1. Consider the scalar equation $x' = a(t)x$, where $a(t)$ is continuous and ω-periodic on $\mathbb{R}$. Solving the IVP

$$\begin{cases} x' = a(t)x, \\ x(0) = x_0, \end{cases}$$

we find $x(t) = x_0 e^{\int_0^t a(s)ds}$.

Therefore, $x(t)$ is ω-periodic iff $x(t + \omega) = x(t)$ iff $x_0 e^{\int_0^{t+\omega} a(s)\,ds} = x_0 e^{\int_0^t a(s)\,ds}$.

If $x_0 \neq 0$, then

$$e^{\int_0^{t+\omega} a(s)\,ds} = e^{\int_0^t a(s)\,ds} \iff e^{\int_t^{t+\omega} a(s)\,ds} = 1$$

$$\iff \int_t^{t+\omega} a(s)\,ds = 0$$

$$\iff \int_0^\omega a(s)\,ds = 0.$$

Example 6.2 (Hill's equation). Consider $x'' + p(t)x = 0$, where $p(t)$ is continuous and ω-periodic on $\mathbb{R}$. Then the corresponding first order system is given by $y' = A(t)y$, where

$$y = \begin{bmatrix} x \\ x' \end{bmatrix}, \quad A(t) = \begin{bmatrix} 0 & 1 \\ -p(t) & 0 \end{bmatrix}.$$

Let $x_1(t)$, $x_2(t)$ be solutions of $x'' + p(t)x = 0$ satisfying the initial condition $x_1(0) = 1$, $x_1'(0) = 0$ and $x_2(0) = 0$, $x_2'(0) = 1$. Then a fundamental matrix solution of $X' = A(t)X$ is given by

$$X(t) = \begin{bmatrix} x_1 & x_2 \\ x_1' & x_2' \end{bmatrix}.$$

By the Floquet theory, we have $X(t + \omega) \equiv X(t)C$, for all t. At $t = 0$, $X(\omega) = X(0)C = IC = C$, so

$$C = \begin{bmatrix} x_1(\omega) & x_2(\omega) \\ x_1'(\omega) & x_2'(\omega) \end{bmatrix}.$$

Notice also that $\det X(t) = \det X(0)e^{\int_0^t 0\, ds} = 1$ by Theorem 5.4, and

$$\det[C - \sigma I] = \begin{vmatrix} x_1(\omega) - \sigma & x_2(\omega) \\ x_1'(\omega) & x_2'(\omega) - \sigma \end{vmatrix}$$
$$= \sigma^2 - (x_1(\omega) + x_2'(\omega))\sigma + 1$$
$$= \sigma^2 - 2\alpha\sigma + 1,$$

where $\alpha = \frac{1}{2}[x_1(\omega) + x_2'(\omega)]$. If $\alpha \in \mathbb{R}$ and $\alpha^2 < 1$, then eigenvalues are $\sigma = \frac{2\alpha \pm \sqrt{4\alpha^2 - 4}}{2} = \alpha \pm i\sqrt{1 - \alpha^2}$, implying $|\sigma|^2 = 1$. So, both eigenvalues satisfy $|\sigma| = 1$, and in this case, from Remark 6.5, all the solutions of the D.E. are bounded on $\mathbb{R}$.

Next, we provide some concluding remarks concerning Hill's equation:

$$x'' + p(t)x = 0, \quad \text{or} \quad x'' + (a + \varphi(t))x = 0,$$

where $\varphi(t)$ is ω-periodic and is a constant.

Remark 6.6. We have that the characteristic multipliers σ_1, σ_2 are roots of $\sigma^2 - 2\alpha\sigma + 1 = 0$, where $2\alpha = x_1(\omega) + x_2'(\omega)$. Since the constant term in the quadratic is 1, we have $\sigma_1\sigma_2 = 1$, and hence $|\sigma_1||\sigma_2| = 1$. If σ_1 and σ_2 are distinct such that $|\sigma_1| = |\sigma_2| = 1$, then all the solutions of the D.E. are bounded on $\mathbb{R}$. (Therefore, the first derivatives are also bounded).

In particular, this will be the case if $\alpha \in \mathbb{R}$ and $\alpha^2 < 1$. On the other hand, if $|\sigma_1| < 1$, then $|\sigma_2| > 1$ and the D.E. has two L.I. solutions, one bounded on $[0, \infty)$ with $\|x(t)\| \to 0$, as $t \to \infty$, and the other unbounded on $[0, +\infty)$.

That is, the solutions

$$\vec{x}_1(t) = \begin{bmatrix} x_1 \\ x_1' \end{bmatrix} = X(t)c_1 \text{ is bounded}$$

and

$$\vec{x}_2(t) = \begin{bmatrix} x_2 \\ x_2' \end{bmatrix} = X(t)c_2 \text{ is unbounded,}$$

where $Cc_1 = \sigma_1 c_1$, $Cc_2 = \sigma_2 c_2$. Hence,

$$\|\vec{x}_1(t + n\omega)\| = |\vec{x}_1 \sigma_1^n| \,\|(t)\| \to 0, \text{ and}$$
$$\|\vec{x}_2(t + n\omega)\| = |\vec{x}_2 \sigma_2^n| \,\|(t)\| \to +\infty.$$

As a last result for the homogeneous equation $X' = A(t)X$, where $A(t)$ is ω-periodic, we have the following remark.

Remark 6.7. $x(t)$ is a nontrivial ω-periodic solution of $x' = A(t)x$ iff 1 is a characteristic multiplier.

Proof. Let $X(t)$ be the solution of

$$\begin{cases} X' = A(t)X, \\ X(0) = I. \end{cases}$$

Assume $x(t)$ is a nontrivial ω-periodic solution of $x' = A(t)x$. Since $x(t)$ is nontrivial, $x(0) = c_0 \neq 0$. Now $x(t) = X(t)c_0$, so $X(t)c_0 = x(t) = x(t + \omega) = X(t + \omega)c_0 = X(t)Cc_0$ (where $X(t + \omega) = X(t)C$). Hence, $Cc_0 = c_0$ which implies 1 is a characteristic multiplier.

For the converse, if 1 is a characteristic multiplier, then $Cc_0 = C_0$, where we again get $X(t + \omega) = X(t)C$. Then $x(t) = X(t)c_0$ will be a nontrivial ω-periodic solution of

$$\begin{cases} X' = A(t)X, \\ X(0) = c_0 \neq 0. \end{cases} \qquad \square$$

6.2   Periodic Nonhomogeneous Linear Systems and Floquet Theory

Let us now turn our consideration to periodic *nonhomogeneous* linear systems. So, consider

$$x' = A(t)x + f(t), \tag{6.1}$$

where $A(t)$ is a continuous $n \times n$ matrix function on $\mathbb{R}$ with period ω, and $f(t)$ is a continuous n-vector function on $\mathbb{R}$ with period ω.

Theorem 6.2. *Let* (6.1) *be a periodic system as defined above. Then a solution* $x(t)$ *of* (6.1) *has period* ω *iff* $x(0) = x(\omega)$.

Proof. Assume $x(t)$ is an ω-periodic solution. Then, $x(t + \omega) = x(t)$, for all $t \in \mathbb{R}$. Hence, $x(\omega) = x(0)$.

On the other hand, assume that $x(t)$ is a solution of (6.1) with $x(0) = x(\omega)$. Then

$$\frac{d}{dt} x(t + \omega) = x'(t + \omega)$$

$$= A(t + \omega)x(t + \omega) + f(t + \omega)$$

$$= A(t)x(t + \omega) + f(t).$$

So, $x(t + \omega)$ is a solution of (6.1). Moreover, $x(t + \omega)|_{t=0} = x(t)|_{t=0}$ by assumption, and hence, $x(t)$ and $x(t + \omega)$ are both solutions of the IVP

$$\begin{cases} x' = A(t)x + f(t), \\ x(0) = x(\omega). \end{cases}$$

By uniqueness of solutions of the IVP, $x(t) \equiv x(t + \omega)$. $\qquad\square$

Corollary 6.1. *Let $A(t)$ be a continuous $n \times n$ matrix function on $\mathbb{R}$ having period ω. Then (6.1) has a unique solution with period ω corresponding to each continuous n-vector $f(t)$ on $\mathbb{R}$ having period ω iff "1" is **not** a characteristic multiplier of $x' = A(t)x$ (iff the only periodic solution with period ω of $x' = A(t)x$ is $x(t) \equiv 0$, by Remark 6.7).*

Proof. First, suppose that (6.1) has a unique solution with the period ω corresponding to each continuous ω-periodic n-vector $f(x)$. Well, $f(t) \equiv 0$ is such a vector. By Remark 6.7, 1 is not a characteristic multiplier of $x' = A(t)x$, because $x(t) \equiv 0$ is the unique ω-periodic solution.

For the other part, assume 1 is not a characteristic multiplier of $x' = A(t)x$. Then, if $X(t)$ is the solution of

$$\begin{cases} X' = A(t)X, \\ X(0) = I, \end{cases}$$

every solution of (6.1) can be written in the form of $x(t) = X(t)x_0 + \int_0^t X(t, s)f(s)\, ds$, where $x(0) = x_0$.

Now (6.1) has an ω-periodic solution iff $x(0) = x(\omega)$ by Theorem 6.2. For $x(0) = x(\omega)$, we must have

$$x_0 = x(0) = x(\omega) = X(\omega)x_0 + \int_0^\omega X(\omega, s)f(s)\, ds.$$

Since 1 is not a characteristic multiplier and since $X(\omega) = C$, we have $X(\omega) - I$ is nonsingular. Therefore x_0 must have the unique value

$$x_0 = -(X(\omega) - I)^{-1} \int_0^\omega X(\omega, s)f(s)\, ds.$$

Then the unique solution $x(t)$ is given by

$$x(t) = -X(t)[X(\omega) - I]^{-1} \int_0^\omega X(\omega, s)f(s)\,ds + \int_0^t X(t, s)f(s)\,ds.$$

Let's examine the solution $x(t)$ in the corollary more closely.

$$x(t) = -X(t)[X(\omega) - I]^{-1} \int_0^\omega X(\omega, s)f(s)\,ds + \int_0^t X(t, s)f(s)\,ds$$
$$= \int_0^\omega G(t, s)f(s)\,ds,$$

where

$$G(t, s) = \begin{cases} -X(t)[X(\omega) - I]^{-1}X(\omega, s) + X(t, s), & 0 \le s < t, \\ -X(t)[X(\omega) - I]^{-1}X(\omega, s), & t \le s \le \omega. \end{cases}$$

Recall that $X(t_1, t_2) = X(t_1)X^{-1}(t_2)$. Hence,

$$G(t, s) = \begin{cases} -X(t)[X(\omega) - I]^{-1}X(\omega)X^{-1}(s) + X(t)X^{-1}(s), & 0 \le s < t, \\ -X(t)[X(\omega) - I]^{-1}X(\omega)X^{-1}(s), & t \le s \le \omega. \end{cases}$$

We can show that

$$G(t, s) = \begin{cases} -X(t)[X(\omega) - I]^{-1}X^{-1}(s), & 0 \le s < t, \\ -X(t)[I + (X(\omega) - I)^{-1}]X^{-1}(s), & t \le s \le \omega. \end{cases} \qquad \square$$

In summary, (6.1) with $A(t)$ continuous and ω-periodic has a unique solution ω-periodic corresponding to each continuous ω-periodic $f(t)$ iff 1 is not an eigenvalue of $X(\omega) = C$, where $X(t)$ is the solution of

$$\begin{cases} X' = A(t)X, \\ X(0) = I, \end{cases}$$

and this unique solution can be expressed as

$$x(t) = \int_0^\omega G(t, s)f(s)ds,$$

where

$$G(t, s) = \begin{cases} -X(t)[X(\omega) - I]^{-1}X^{-1}(s), & 0 \le s < t, \\ -X(t)[I + (X(\omega) - I)^{-1}]X^{-1}(s), & t \le s \le \omega. \end{cases}$$

Here, $G(t, s)$ is called a *Green's function*.

Example 6.3. Consider the scalar equation $x' = a(t)x + f(t)$, $a(t)$ and $f(t)$ are continuous on $\mathbb{R}$ and ω-periodic. Find the Green's function. Here $A(t) = (a(t))_{1 \times 1}$ and $(a(t))^{-1} = \left(\frac{1}{a(t)}\right)$. The solution $x(t)$ of

$$\begin{cases} x' = a(t)x, \\ x(0) = 1 \end{cases}$$

is given by $x(t) = e^{\int_o^t a(r)\, dr}$.

Now, if 1 is not an eigenvalue of $x(\omega)$, then $x(\omega) \neq 1$; i.e., $\int_0^\omega a(r)\, dr \neq 0$. So in the case, $x' = a(t)x + f(x)$ has a periodic solution for each ω-periodic $f(t)$ given by $x(t) = \int_0^\omega G(t, s) f(s)\, ds$ where

$$G(t, s) = \begin{cases} \dfrac{e^{\int_s^t a(r)\, dr}}{1 - e^{\int_0^\omega a(r)\, dr}}, & 0 \leq s < t, \\[3mm] -e^{\int_s^t a(r)\, dr} + \dfrac{e^{\int_s^t a(r)\, dr}}{1 - e^{\int_0^\omega a(r)\, dr}}, & t \leq s \leq \omega. \end{cases}$$

In considering a Green's function, one can fix s and consider t-values on both sides of s, or one can fix t and consider s-values on both sides of t.

Let us look at some properties of the above Green's function, $G(t, s)$.

(1) For a fixed s, $0 < s < \omega$, $G(t, s)$ as a function of t is a solution of $x' = a(t)x$ on $[0, s]$ and on $[s, \omega]$; e.g., for $[0, s]$ and $t \leq s$, $-e^{\int_s^t a(r)\, dr}$ is a solution of $x' = a(t)x$ and any linear combination is also a solution. Thus,

$$-e^{\int_s^t a(r)dr} + \frac{e^{\int_s^t a(r)dr}}{1 - e^{\int_0^\omega a(r)dr}}$$

is a solution on $[0, s]$. Similarly on $[s, \omega]$ and $t \geq s$,

$$\frac{e^{\int_s^t a(r)dr}}{1 - e^{\int_0^\omega a(r)dr}}$$

is a solution. Hence, $G(t, s)$ is a solution of $x' = a(t)x$ on $[0, s]$ and on $[s, \omega]$.

(2) For s fixed with $0 < s < \omega$,

$$G(0, s) = -e^{\int_s^0 a(r)dr} + \frac{e^{\int_s^0 a(r)dr}}{1 - e^{\int_0^\omega a(r)dr}}$$

$$= -e^{-\int_0^s a(r)dr} + \frac{e^{-\int_0^s a(r)dr}}{1 - e^{\int_0^\omega a(r)dr}}$$

$$= \frac{e^{\int_s^\omega a(r)dr}}{1 - e^{\int_0^\omega a(r)dr}} = G(\omega, s),$$

i.e., $G(0, s) = G(\omega, s)$.

(3) Fix s with $0 < s < \omega$, and let t approach s from the right (denoted s^+) and also let t approach s from the left (denoted s^-). Then

$$G(s^+, s) - G(s^-, s) = +1.$$

In this example, since we are assuming that $e^{\int_0^\omega a(r)dr} \neq 1$, it follows that <u>no</u> solution $x(t)$ of

$$\begin{cases} x' = a(t)x, \\ x(0) = x_0 \end{cases}$$

is a nontrivial periodic solution by the corollary. Yet, the Green's function is *almost* a periodic solution of $x' = a(t)x$, but from (3) above, it has a jump of $+1$ at s. See Figure 6.2.

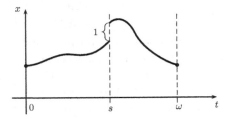

Fig. 6.2 $G(t,s)$ has a jump of $+1$ at $t = s$.

We would next like to show the existence of solutions which are periodic using a fixed point theorem. One of the interesting asides of this fixed point theorem is that we can also apply it to an alternate proof of the Picard Existence Theorem.

Theorem 6.3 (Contraction Mapping Principle). *Let $\langle M, d \rangle$ be a complete metric space and assume that there exist a function $T : M \to M$ and a constant K, $0 \leq K < 1$, such that $d(T(x), T(y)) \leq Kd(x,y)$, for each x, $y \in M$. Then T has a unique fixed point; i.e., there exists a unique $x_0 \in M$ such that $T(x_0) = x_0$.*

Proof. Clearly, T is continuous. Now let $z \in M$ be arbitrarily chosen and form the sequence, $z, Tz, T^2z = T(Tz), T^3z, \ldots$. Then,

$$d(T^{n+1}(z), T^n(z)) = d(T(T^n(z)), T(T^{n-1}(z)))$$
$$\leq Kd(T^n(z), T^{n-1}(z))$$
$$\vdots$$
$$\leq K^n d(T(z), z).$$

Assume $n > m > 1$, then

$$d(T^n(z), T^m(z))$$
$$\leq d(T^n(z), T^{n-1}(z)) + d(T^{n-1}(z), T^{n-2}(z))$$
$$+ \cdots + d(T^{m+1}(z), T^m(z))$$
$$\leq K^{n-1}d(T(z), z) + K^{n-2}d(T(z), z) + \cdots + K^m d(T(z), z)$$
$$= (K^{n-1} + K^{n-2} + \cdots + K^m)d(T(z), z)$$
$$\leq \sum_{j=m}^{\infty} K^j d(T(z), z)$$
$$= \frac{K^m}{1 - K}d(T(z), z).$$

Then, as $m, n \to \infty$, $d(T^n(z), T^m(z)) \to 0$, hence $\{T^n(z)\}_{n=0}^{\infty}$ is a Cauchy sequence. M is complete, thus there exists $x_0 \in M$ such that $\lim_{n \to \infty} T^n(z) = x_0$. Since T is continuous,

$$T(x_0) = T\left(\lim_{n \to \infty} T^n(z)\right) = \lim_{n \to \infty} T^{n+1}(z) = x_0.$$

So x_0 is "a" fixed point of T. For the uniqueness, let y_0 also be a fixed point of T. Then

$$d(x_0, y_0) = d(T(x_0), T(y_0)) \leq Kd(x_0, y_0),$$

i.e., $(1 - K)d(x_0, y_0) \leq 0$. Since $(1 - K) > 0$, we have $d(x_0, y_0) = 0$. Hence, $x_0 = y_0$. So, T has a unique fixed point x_0. $\square$

Corollary 6.2. *Assume the conditions of the Contraction Mapping Principle, except assume that for some $m \geq 1$, T^m is a contraction; i.e., there exists $0 \leq K < 1$ such that $d(T^m(x), T^m(y)) \leq K(x, y)$, for all $x, y \in M$. Then T has a unique fixed point.*

Proof. In this case we consider the sequence $\{T^{jm}(z)\}_{j=0}^{\infty}$. By the Contraction Mapping Principle, T^m has a unique fixed point $x_0 \in M$, that is, $T^m(x_0) = x_0$. Then $T(T^m(x_0)) = T^m(T(x_0)) = T(x_0)$. So, $T(x_0)$ is a fixed point of T^m, and so by uniqueness of x_0, we have $T(x_0) = x_0$. Hence, x_0 is a fixed point of T.

If $Ty_0 = y_0$, then we have $T^m(y_0) = y_0$ and $T^m(x_0) = x_0$ and so $y_0 = x_0$. Thus, T has a unique fixed point x_0. $\qquad\square$

Before presenting our alternate proof of the Picard Existence Theorem, we define a metric on the set of continuous n-vector-valued functions on $[a, b]$.

Let the metric space

$$M = \{x(t) \mid x(t) \text{ is a continuous } n\text{-vector function on } [a, b]\}$$

with metric

$$d(x, y) \equiv \max_{t \in [a,b]} \|x(t) - y(t)\| \equiv \|x - y\|.$$

That $\langle M, d \rangle$ is a complete metric space follows from an application of the Arzelà-Ascoli Theorem.

Theorem 6.4 (Picard Existence Theorem). *Assume $f(t, x) : [a, b] \times \mathbb{R}^n \to \mathbb{R}^n$ is continuous and satisfies a Lipschitz condition $\|f(t, x) - f(t, y)\| \le K\|x - y\|$, for all $(t, x), (t, y) \in [a, b] \times \mathbb{R}^n$, where $K > 0$. Then given $(t_0, x_0) \in [a, b] \times \mathbb{R}^n$, the IVP*

$$\begin{cases} x' = f(t, x), \\ x(t_0) = x_0, \end{cases}$$

has a unique solution on $[a, b]$.

Proof. Define $T : M \to M$ by

$$(Tx)(t) = x_0 + \int_{t_0}^{t} f(s, x(s)) \, ds.$$

We claim that T has a unique fixed point.

Consider

$$\|(Tx)(t) - (Ty)(t)\| \le K \left| \int_{t_0}^{t} \|x(s) - y(s)\| \, ds \right|$$

$$\le K \left| \int_{t_0}^{t} \max_{s \in [a,b]} \|x(s) - y(s)\| \, ds \right|$$

$$= K \, |t - t_0| \, \|x - y\|.$$

Hence,

$$\|Tx - Ty\| \le K(b - a)\|x - y\|.$$

Now

$$(T^2 x)(t) = T(Tx)(t) = x_0 + \int_{t_0}^t f(s, (Tx(s))) \, ds,$$

and so,

$$\left\| (T^2 x)(t) - (T^2 y)(t) \right\| \leq K \left| \int_{t_0}^t \| (Tx)(s) - (Ty)(s) \| \, ds \right|$$

$$\leq K^2 \| x - y \| \int_{t_0}^t |s - t_0| \, ds$$

$$= K^2 \| x - y \| \frac{|t - t_0|^2}{2},$$

$$\vdots$$

$$\| (T^m x)(t) - (T^m y)(t) \| \leq \frac{K^m |t - t_0|^m}{m!} \| x - y \| \leq \frac{K^m (b - a)^m}{m!} \| x - y \|.$$

So,

$$\| T^m x - T^m y \| \leq \frac{K^m (b - a)^m}{m!} \| x - y \|.$$

Now for m sufficiently large, we have $\frac{K^m (b-a)^m}{m!} < 1$. Hence, for large m, T^m is a contraction and so by Corollary 6.2, T has a unique fixed point $u(t) \in M$. So,

$$u(t) = (Tu)(t) = x_0 + \int_{t_0}^t f(s, u(s)) ds$$

which is equivalent to the statement that $u(t)$ is a solution of

$$\begin{cases} u' = f(t, u), \\ u(t_0) = x_0, \end{cases}$$

and this solution is unique since u is the unique fixed point. □

For our last result in this section, we are concerned with periodic solutions for the scalar D.E. $x' = a(t)x + f(t, x)$. We will apply the Contraction Mapping Principle. For this, let

$$M \equiv \{ \text{All } \omega\text{-periodic continuous functions on } \mathbb{R} \}$$

with metric $d(\phi, \psi) = \max_{t \in \mathbb{R}} |\phi(t) - \psi(t)| = \max_{0 \leq t \leq w} |\phi(t) - \psi(t)|$. Then $\langle M, d \rangle$ is a complete metric space.

Theorem 6.5. *Let $a(t) \in M$ and let $f(t, x)$ be continuous on $\mathbb{R} \times \mathbb{R}$ with $f(t, x)$ ω-periodic in t, for each fixed x (Note that for $\phi \in M$, $f(t, \phi(t))$ is*

ω-*periodic.) Assume* $\int_0^\omega a(r)\,dr \neq 0$, $|f(t,x) - f(t,y)| \leq K|x - y|$, *for all* $(t,x),(t,y) \in \mathbb{R} \times \mathbb{R}$, *and* $\max_{0 \leq t \leq w} \int_0^\omega |G(t,s)|\,ds < \frac{1}{K}$. *Then the D.E.* $x' = a(t)x + f(t,x)$ *has a unique* ω-*periodic solution.*

Proof. If $\int_0^\omega a(r)\,dr \neq 0$, then $x' = a(t)x$ does not have nontrivial ω-periodic solution. Hence, we can form a Green's function $G(t,s)$.

Now define $T : M \to M$ via

$$(T\phi)(t) = \int_0^\omega G(t,s)f(s,\phi(s))\,ds.$$

Consider

$$|(T\phi)(t) - (T\psi)(t)| \leq \int_0^\omega |G(t,s)|K|\phi(s) - \psi(s)|\,ds$$

$$\leq d(\phi,\psi)\int_0^\omega |G(t,s)|K\,ds.$$

So by the Contraction Mapping Principle, T has a unique fixed point ϕ such that

$$\phi(t) = (T\phi)(t) = \int_0^\omega G(t,s)f(s,\phi(s))\,ds,$$

which is the unique ω-periodic solution of $x' = a(t)x + f(t,x)$. $\square$

Exercise **36.** Calculate $\max_{t \in [0,\omega]} \int_0^\omega |G(t,s)|\,ds$ in the case $a(t) \equiv a$ (constant) which has any desired period.

Chapter 7

Stability Theory

7.1 Stability of First Order Systems

We will assume that in the following definitions, we are talking about a fixed first order system $x' = f(t, x)$ in which $f(t, x)$ is continuous on $[t_0, +\infty) \times \mathbb{R}^n$ or continuous on $[t_0, \infty) \times B(0, R)$, where $B(0, R) = \{x \in \mathbb{C}^n \mid \|x\| < R\}$ (sometimes denoted $B_R(0)$).

Definition 7.1. The solution $x(t; t_0, x_0)$ is said to be *stable* on $[t_0, \infty)$ in case for each $\varepsilon > 0$, there exists $\delta > 0$ such that $\|x_1 - x_0\| < \delta$ implies that all solutions $x(t; t_0, x_1)$ exist on $[t_0, \infty)$ and $\|x(t; t_0, x_1) - x(t; t_0, x_0)\| < \varepsilon$ on $[t_0, \infty)$.

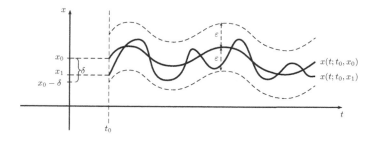

Fig. 7.1   Stable solution.

Example 7.1 (Stability).

(1) $x' = 0$. Let $x(t; 0, 0)$ be a solution on $[0, \infty)$. Then, $x(t; 0, 0) \equiv 0$. This is clearly stable by uniqueness of solutions and the nature of the equation $x' = 0$.

But to verify the stability of $x(t; 0, 0) \equiv 0$, let $\delta = \varepsilon$. Then if $\|x_1 - 0\| < \delta$, the solution of $x' = 0$, $x(0) = x_1$ is given by $x(t; 0, x_1) = x_1$ and

$$\|x(t; 0, x_1) - x(t; 0, 0)\| = \|x_1 - 0\| < \delta = \varepsilon.$$

(2) $x' = -x$ on $[0, \infty)$. In this case $x(t; 0, 0) \equiv 0$. Let $x(t; 0, x_1)$ be any other solution satisfying $x(0) = x_1$. Then $x(t; 0, x_1) = x_1 e^{-t}$. Consider $|x(t; 0, x_1) - x(t; 0, 0)| = |x_1| e^{-t} \leq |x_1|$. Thus taking $\delta = \varepsilon$, and the solution $x(t; 0, 0)$ is stable on $[0, \infty)$.

(3) $x' = x$ on $[0, \infty)$. Again, $x(t; 0, 0) \equiv 0$. For any x_1, $x(t; 0, x_1) = x_1 e^{t}$ and so $|x(t; 0, x_1) - x(t; 0, 0)| = |x_1| e^{t} \to \infty$ as $t \to +\infty$. So, $x(t; 0, 0) \equiv 0$ is unstable on $[0, \infty)$.

Definition 7.2. The solution $x(t; t_0, x_0)$ is said to be *asymptotically stable* on $[t_0, \infty)$ in case

(1) $x(t; t_0, x_0)$ is stable on $[t_0, \infty)$, and

(2) there exists $\eta > 0$ such that $\|x_1 - x_0\| < \eta$ implies

$$\lim_{t \to \infty} \|x(t; t_0, x_1) - x(t, t_0, x_0)\| = 0,$$

i.e., the solutions approach each other asymptotically.

Example 7.2. In Example 7.1(2) for $x' = -x$, the solution $x(t; 0, 0) \equiv 0$ is asymptotically stable because $|x_1| e^{-t} \to 0$, as $t \to \infty$.

Note: The above definitions of stability and asymptotic stability are due to Lyapunov.

Definition 7.3. The solution $x(t; t_0, x_0)$ is said to be *uniformly stable* on $[t_0, \infty)$ in case for each $\varepsilon > 0$, there exists $\delta_\varepsilon > 0$ such that if $t_1 \geq t_0$ and $\|x_1 - x(t_1; t_0, x_0)\| < \delta_\varepsilon$, then $\|x(t; t_1, x_1) - x(t; t_0, x_0)\| < \varepsilon$ on $[t_1, \infty)$.

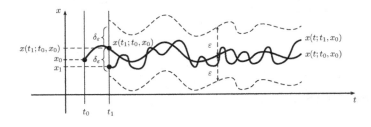

Fig. 7.2   Uniformly stable solution.

Definition 7.4. The solution $x(t; t_0, x_0)$ is said to be *uniformly asymptotically stable* on $[t_0, \infty)$ in case

(1) $x(t; t_0, x_0)$ is uniformly stable on $[t_0, \infty)$;

(2) there exists $\delta_0 > 0$ such that $t_1 \geq t_0$ and $\|x_1 - x(t_1; t_0, x_0)\| < \delta_0$, imply $\lim_{t \to \infty} \|x(t; t_1, x_1) - x(t; t_0, x_0)\| = 0$ (i.e., solutions are squeezed together past t_1), and

(3) for each $\varepsilon > 0$, there exists $T(\varepsilon) > 0$ such that $t_1 \geq t_0$ and $\|x_1 - x(t_1; t_0, x_0)\| < \delta_0$ imply $\|x(t; t_1, x_1) - x(t; t_0, x_0)\| < \varepsilon$, for $t \geq t_1 + T(\varepsilon)$ (this says the uniformity of the squeezing is bounded by ε, for any t past $t_1 + T(\varepsilon)$).

Definition 7.5. The solution $x(t; t_0, x_0)$ is said to be *strongly stable* on $[t_0, \infty)$ in case for each $\varepsilon > 0$, there exists $\delta_\varepsilon > 0$ such that $t_1 \geq t_0$ and $\|x_1 - x(t_1; t_0, x_0)\| < \delta_\varepsilon$ imply $x(t; t_1, x_1)$ exists on $[t_0, \infty)$ and $\|x(t; t_1, x_1) - x(t; t_0, x_0)\| < \varepsilon$ on $[t_0, \infty)$.

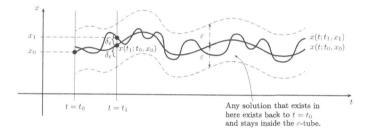

Fig. 7.3   Strongly stable solution.

Obviously,

Strong Stability $\Rightarrow$ Uniform Stability $\Rightarrow$ Stability;

Uniform Asymptotic Stability $\Rightarrow$ Asymptotic Stability.

| Exercise | **37.** Show that the following definition of stability is equivalent to the given one if solutions of IVP's for $x' = f(t, x)$ are unique: $x(t; t_0, x_0)$ is *stable* on $[t_0, \infty)$ in case given $t_1 \geq t_0$ and given $\varepsilon > 0$, there exists $\delta(t_1, \varepsilon) > 0$ such that $\|x_1 - x(t_1; t_0, x_0)\| < \delta$ implies: $x(t; t_1, x_1)$ exists on $[t_1, \infty)$ and $\|x(t; t_1, x_1) - x(t; t_0, x_0)\| < \varepsilon$ on $[t_1, \infty)$.

Note: This clearly implies Definition 7.1 by taking $t_1 = t_0$. Hint for the converse: Let x_n be a sequence of tuples with $x_n \to x(t_1; t_0, x_0)$ and consider the sequence of solutions $x(t; t_1, x_n)$. Applying the Kamke Theorem, the solutions approach $x(t; t_0, x_0)$ on compact subintervals of $[t_0, \infty)$. Then apply Definition 7.1 and the stated property in the exercise will be satisfied.

Note then that our Definition 7.1 and the statement in Exercise 37 are not equivalent, if solutions of IVP's are not unique to both the right and the left. This says that our Definition 7.1 of stability gives uniqueness of solutions to the right but not necessarily to the left.

Example 7.3 (Uniformly Asymptotically Stability). Consider the solution $x(t) \equiv 0$ of $x' = -x$ on $[0, +\infty)$.

(1) Any solution is of the form $x(t; t_1, x_1) = x_1 e^{-(t-t_1)}$. Hence $|x(t; t_1, x_1) - x(t; 0, 0)| = |x_1| e^{-(t-t_1)} \leq |x_1|$ on $[t_1, \infty)$. Thus, taking $\delta = \varepsilon$, $x(t) \equiv 0$ is uniformly stable.

(2) Let δ_0 be a fixed positive number; then $|x_1 - x(t; 0, 0)| = |x_1| < \delta_0$ implies $\lim_{t \to \infty} |x(t; t_1, x_1) - x(t; 0, 0)| = \lim_{t \to \infty} |x_1| e^{-(t-t_1)} = 0$. Thus, any $\delta_0 > 0$ works for this part.

(3) For this part, we proceed to find $T(\varepsilon)$. Now $|x(t; t_1, x_1) - x(t; 0, 0)| = |x_1| e^{-(t-t_1)} < \delta_0 e^{-(t-t_1)}$, where $|x_1| < \delta_0$. (For $t \geq t_1 + T(\varepsilon)$, take $\varepsilon = \delta_0 e^{-(t-t_1)}$. Then, $t - t_1 = \ln \frac{\delta_0}{\varepsilon}$.) Take $T(\varepsilon) = \ln \frac{\delta_0}{\varepsilon}$, so that if $-(t - t_1) \leq -T(\varepsilon)$, then $e^{-(t-t_1)} \leq e^{-T(\varepsilon)} = \frac{\varepsilon}{\delta_0}$. Therefore, for $t \geq t_1 + \ln \frac{\delta_0}{\varepsilon}$,

$$|x(t; t_1, x_1) - x(t; 0, 0)| = |x_1| e^{-(t-t_1)} < \delta_0 \left(\frac{\varepsilon}{\delta_0} \right) = \varepsilon.$$

Thus, $x(t) \equiv 0$ is uniformly asymptotically stable.

$\boxed{\text{Exercise}}$ **38.** Show that the solution $x(t) \equiv 0$ of $x'' + x = 0$ is strongly stable on $[0, \infty)$.

Convert to

$$\begin{bmatrix} x \\ x' \end{bmatrix}' = \begin{bmatrix} 0 & 1 \\ -1 & 0 \end{bmatrix} \begin{bmatrix} x \\ x' \end{bmatrix}$$

and then show that not only are the x-values close to 0, but so also are the x'-values; i.e.,

$$\left\| \begin{bmatrix} x \\ x' \end{bmatrix} \right\| = \max \{ |x(t)|, |x'(t)| \}$$

are close to 0.

Example 7.4. Consider the solution $x(t) \equiv 0$ of the scalar equation $x' = a(t)x$. Now $x(t; t_1, x_1) = x_1 e^{\int_{t_1}^t a(s)\,ds}$.

(1) $x(t) \equiv 0$ is stable on $[0, \infty)$ iff Re $\int_0^t a(s)\,ds \leq M$(constant) on $[0, \infty)$. To see this, take

$$|x(t; 0, x_1) - x(t; 0, 0)| = |x_1| e^{\text{Re}\int_0^t a(s)\,ds} < \varepsilon,$$

iff $e^{\text{Re}\int_0^t a(s)\,ds}$ is bounded, iff Re $\int_0^t a(s)\,ds \leq M$ for some $M > 0$.

(2) $x(t) \equiv 0$ is asymptotically stable on $[0, \infty)$ iff Re $\int_0^t a(s)\,ds \to -\infty$.

(3) $x(t) \equiv 0$ is uniformly stable on $[0, \infty)$ iff there exists $M \geq 0$ such that Re $\int_{t_1}^t a(s)\,ds \leq M$, for all $t \geq t_1 \geq 0$.

Example 7.5 (Asymptotic Stability $\not\Rightarrow$ Uniform Stability). Consider the solution $x(t) \equiv 0$ of $x' = a(t)x$ on $[1, \infty)$, where $a(t) = \sin(\ln t) + \cos(\ln t) - \alpha$, where $1 < \alpha < \sqrt{2}$. Then

$$\int_1^t a(s)\,ds = t\sin(\ln t) - \alpha t + \alpha,$$

so solutions are of the form

$$x(t) = x_0 e^{\int_1^t a(s)\,ds} = x_0 e^{t[\sin(\ln t) - \alpha] + \alpha},$$

where $x(1) = x_0$. Note: $\sin(\ln t) - \alpha < 0$, since $1 < \alpha < \sqrt{2}$.

So, $|x(t)| \leq |x_0| e^{\alpha}$, for $t \geq 1$, since $\alpha > 1$. Thus the zero solution is stable on $[1, \infty)$. Also $\lim_{t \to \infty} |x(t)| = \lim_{t \to \infty} x_0 e^{t(\sin(\ln t) - \alpha) + \alpha} = 0$. Therefore the zero solution is asymptotically stable on $[1, \infty)$.

However, the zero solution is not uniformly stable. To see this, consider $\sin \frac{\pi}{4} + \cos \frac{\pi}{4} = \sqrt{2}$ and $\alpha < \sqrt{2}$. By continuity we can fix $\alpha < \beta < \sqrt{2}$ and an interval $[\theta_1, \theta_2]$ such that $\sin \theta + \cos \theta \geq \beta$ on $[\theta_1, \theta_2]$. See Figure 7.4.

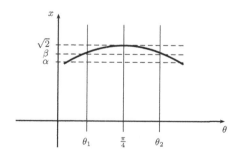

Fig. 7.4 $\sin \theta + \cos \theta \geq \beta$ on $[\theta_1, \theta_2]$.

For $n = 1, 2, \ldots$, let $t_{1n} = e^{2n\pi + \theta_1}$ and $t_{2n} = e^{2n\pi + \theta_2}$. Then $\sin(\ln t) + \cos(\ln t) \geq \beta$ on $[t_{1n}, t_{2n}]$. Hence $\int_{t_{1n}}^{t_{2n}} a(s)\, ds \geq \int_{t_{1n}}^{t_{2n}} (\beta - \alpha)\, ds$, because $a(t) \geq \beta - \alpha$. Therefore, as $n \to \infty$,

$$\int_{t_{1n}}^{t_{2n}} a(s)\, ds \geq (\beta - \alpha)(t_{2n} - t_{1n}) = e^{2n\pi}(\beta - \alpha)(e^{\theta_2} - e^{\theta_1}) \to +\infty.$$

This can be intuitively seen from Figure 7.5.

Fig. 7.5   Graph of function $a(t)$.

Now if $x(t)$ is a solution in the δ_ε-tube at t_{1n}, it does not stay in the ε-tube. In fact, let $x(t)$ be a solution such that $x(t_{1n}) = \frac{\delta_\varepsilon}{2}$, then $x(t) = \frac{\delta_\varepsilon}{2} e^{\int_{t_{1n}}^{t} a(s)\, ds}$ for $t \geq t_{1n}$. But from above, $x(t_{2n}) \geq \frac{\delta_\varepsilon}{2} \exp\left(e^{2n\pi}(\beta - \alpha)(e^{\theta_2} - e^{\theta_1})\right)$ which can be made arbitrarily large for n large enough. In particular $|x(t_{2n})| \not< \varepsilon$ for large n and hence the zero solution is not uniformly stable. See Figure 7.6.

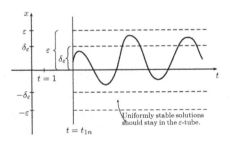

Fig. 7.6

7.2   Stability of Vector Linear Systems

For our next consideration, we discuss stability concepts for a vector linear system

$$x' = A(t)x + f(t), \tag{7.1}$$

where $A(t)$ and $f(t)$ are continuous on $[t_0, \infty)$. Let $x(t; t_0, x_0)$ be a solution of (7.1). This solution is *stable* on $[t_0, \infty)$ in case for each $\varepsilon > 0$, there exists

$\delta_\varepsilon > 0$ such that $\|x_1 - x_0\| < \delta$ implies $\|x(t; t_0, x_1) - x(t; t_0, x_0)\| < \varepsilon$ on $[t_0, \infty)$.

Let

$$y' = A(t)y \tag{7.2}$$

be the homogeneous system and let "$x(t)$" and "$y(t)$" denote solutions of (7.1) and (7.2) respectively.

Note that

$$y(t; t_0, x_1 - x_0) = x(t; t_0, x_1) - x(t; t_0, x_0).$$

Conversely,

$$x(t; t_0, x_0) + y(t; t_0, y_0) = x(t; t_0, x_0 + y_0).$$

Remark 7.1. The solution $x(t; t_0, x_0)$ of (7.1) is stable on $[t_0, \infty)$ iff the zero solution of (7.2) is stable on $[t_0, \infty)$.

Proof. First, let us note that the zero solution of (7.2) is stable on $[t_0, +\infty)$ in case, given $\varepsilon > 0$, there exists $\delta_\varepsilon > 0$ such that $\|y_1\| < \delta$ implies $\|y(t; t_0, y_1)\| < \varepsilon$ on $[t_0, +\infty)$.

Now let us suppose first that $x(t; t_0, x_0)$ is stable and let $\varepsilon > 0$ be given and δ_ε be the corresponding delta.

Now let y_1 be such that $\|y_1\| < \delta_\varepsilon$. Then $\|y(t; t_0, y_1)\| = \|x(t; t_0, y_1 + x_0) - x(t; t_0, x_0)\| < \varepsilon$ by the stability of $x(t; t_0, x_0)$. Therefore, the zero solution of (7.2) is stable.

The converse is similar. Assume the zero solution of (7.2) is stable with ε and δ_ε as usual. If $\|x_1 - x_0\| < \delta_\varepsilon$, then $\|x(t; t_0, x_1) - x(t; t_0, x_0)\| = \|y(t; t_0, x_1 - x_0)\| < \varepsilon$ by the stability of zero solution of (7.2). So, the solution $x(t; t_0, x_0)$ is stable on $[t_0, \infty)$.

Remark 7.2. The same argument shows that for each of the other four types of stability, a fixed solution of (7.1) has that type of stability iff the zero solution of (7.2) has that same type of stability.

Therefore, for a fixed equation (7.1), *all solutions* have the same type of stability, because of the "iff" of the stability of the zero solution of (7.2), and the type of stability is determined by the homogeneous system (7.2). Hence, we say that the **system** (7.2) is stable, is unstable, is asymptotically stable, etc. $\qquad\qquad\square$

Theorem 7.1. *Let $A(t)$ be a continuous $n \times n$ matrix function on $[t_0, \infty)$, and let $X(t)$ be the solution of*

$$\begin{cases} X' = A(t)X, \\ X(t_0) = I. \end{cases}$$

Then the system (7.2), $x' = A(t)x$, is

(1) *Stable on $[t_0, \infty)$ iff there exists $K > 0$ such that $\|X(t)\| \leq K$ on $[t_0, \infty)$.*

(2) *Uniformly stable on $[t_0, \infty)$ iff there exists $K > 0$ such that $\|X(t)X^{-1}(s)\| \leq K$, for all $t_0 \leq s \leq t$.*

(3) *Strongly stable on $[t_0, \infty)$ iff there exists $K > 0$ such that $\|X(t)\| \leq K$ and $\|X^{-1}(t)\| \leq K$ on $[t_0, \infty)$.*

(4) *Asymptotically stable on $[t_0, \infty)$ iff $\lim_{t \to \infty} \|X(t)\| = 0$.*

(5) *Uniformly asymptotically stable on $[t_0, \infty)$ iff there exist $K > 0$ and $\alpha > 0$ such that $\|X(t)X^{-1}(s)\| \leq Ke^{-\alpha(t-s)}$, on $t_0 \leq s \leq t$.*

Proof. (1) Assume that $\|X(t)\| \leq K$. The solution $x(t; t_0, x_0) = X(t)x_0$, so $\|x(t; t_0, x_0)\| = \|X(t)x_0\| \leq K\|x_0\| < \varepsilon$, provided $\|x_0\| < \delta_\varepsilon = \frac{\varepsilon}{K}$. The zero solution of (7.2) is stable which is what we mean by saying the system (7.2) is stable.

Conversely, assume $x' = A(t)x$ is stable on $[t_0, \infty)$. Then, for $\varepsilon = 1$, there exists $\delta_0 > 0$ such that $\|x_0\| < \delta_0$ implies $\|x(t; t_0, x_0)\| < 1$ on $[t_0, \infty)$. So, $\|X(t)x_0\| < 1$ on $[t_0, \infty)$. Hence, $\|X(t)\left(\frac{x_0}{\delta_0}\right)\| < \frac{1}{\delta_0}$ on $[t_0, \infty)$. Now let y_0 be a vector such that $\|y_0\| < 1$. Let $x_0 = \delta_0 y_0$, then $\|x_0\| < \delta_0$. Hence, $\|X(t)y_0\| = \|X(t)\left(\frac{x_0}{\delta_0}\right)\| < \frac{1}{\delta_0}$ for all $t \in [t_0, \infty)$. By the definition of matrix norm, $\|X(t)\| = \sup_{\|y\| \leq 1} \|X(t)y\| \leq \frac{1}{\delta_0}$ on $[t_0, \infty)$. Taking $K = \frac{1}{\delta_0}$, the converse is satisfied.

(2) Assume now that $\|X(t)X^{-1}(s)\| \leq K$. Let $s \in [t_0, \infty)$ and $\varepsilon > 0$ be given. Then for $t \geq s$, $\|x(t; s, x_0)\| = \|X(t)X^{-1}(s)x_0\| \leq K\|x_0\| < \varepsilon$, for $s \leq t$, provided $\|x_0\| < \delta_\varepsilon = \frac{\varepsilon}{K}$ (which is independent of s). Thus (7.2) is uniformly stable.

Conversely, assume (7.2) is uniformly stable on $[t_0, \infty)$. Then, given $\varepsilon > 0$, there exists $\delta_\varepsilon > 0$ such that $t_1 \geq t_0$ and $\|x_0\| < \delta$ imply $\|x(t; t_1, x_0)\| < \varepsilon$ on $[t_1, \infty)$. Hence, $\|X(t)X^{-1}(t_1)x_0\| < \varepsilon$ on $[t_1, \infty)$. By taking $\varepsilon = 1$, the same argument as above can be used to show that $\|X(t)X^{-1}(t_1)\| < \frac{1}{\delta_0}$ on $[t_0, \infty)$.

Remark 7.3. The system (7.2) is stable on $[t_0, \infty)$ iff each solution of $x' = A(t)x$ is bounded on $[t_0, \infty)$.

(3) Assume $\|X(t)\| \leq K$ and $\|X^{-1}(t)\| \leq K$ on $[t_0, \infty)$. Let $x(t; t_1, x_0)$ be a solution with $x(t_1) = x_0$. Then $\|x(t; t_1, x_0)\| = \|X(t)X^{-1}(t_1)x_0\| \leq K^2\|x_0\| < \varepsilon$, if $\|x_0\| < \delta = \frac{\varepsilon}{K^2}$, on $[t_0, \infty)$. Thus (7.2) is strongly stable on $[t_0, \infty)$.

For the converse, assume (7.2) is strongly stable on $[t_0, \infty)$. Then given any $t_1 \geq t_0$, and any $\varepsilon > 0$, (in particular, take $\varepsilon = 1$), there exists $\delta_0 > 0$ such that $\|x_0\| \leq \delta_0$ implies $\|x(t; t_1, x_0)\| < 1$ on $[t_0, \infty)$. There are two cases:

Case 1: Take $t_1 = t_0$. Then, $\|x(t; t_0, x_0)\| = \|X(t)x_0\| < 1$ on $[t_0, \infty)$ for $\|x_0\| < \delta_0$. Part (1) of *this theorem*, this implies $\|X(t)\| \leq \frac{1}{\delta_0}$.

Case 2: Take $t = t_0$. Then,

$$\|x(t_0; t_1, x_0)\| = \|X(t_0)X^{-1}(t_1)x_0\| = \|X^{-1}(t_1)x_0\| < 1,$$

for $\|x_0\| < \delta_0$. Then, as before $\|X^{-1}(t_1)\| \leq \frac{1}{\delta_0}$. But t_1 is arbitrary, hence, $\|X^{-1}(t)\| \leq \frac{1}{\delta_0}$.

(4) Assume that $\|X(t)\| \to 0$ as $t \to +\infty$. Since $\|X(t)\|$ is continuous on $[t_0, \infty)$, $\|X(t)\|$ is bounded on $[t_0, \infty)$. By part (1) of *this theorem*, the equation (7.2) is stable on $[t_0, \infty)$. Secondly, if $x_0 \in \mathbb{R}^n$, then $\|x(t; t_0, x_0)\| = \|X(t)x_0\| \leq \|X(t)\|\|x_0\|$. Hence, $\lim_{t\to\infty} \|x(t; t_0, x_0)\| = 0$. Therefore (7.2) is asymptotically stable on $[t_0, \infty)$.

For the converse, if (7.2) is asymptotically stable on $[t_0, \infty)$, then (7.2) is stable, hence $\|X(t)\|$ is bounded (see the remark above). Also, by the asymptotically stability of (7.2), there exists $\eta > 0$ such that $\|x_0\| \leq \eta$ implies $\lim_{t\to\infty}\|x(t; t_0, x_0)\| = 0$. But we don't need $\eta > 0$ such that $\|x_0\| \leq \eta$; i.e., given $x_0 \in \mathbb{R}^n$, there exists $K \in \mathbb{R}$ such that $\left\|\frac{x_0}{K}\right\| \leq \eta$, and we also have $x(t; t_0, x_0) = K\,x\left(t; t_0, \frac{x_0}{K}\right)$. Thus $\lim_{t\to\infty}\|x(t; t_0, x_0)\| = |K|\lim_{t\to\infty}\left\|x\left(t; t_0, \frac{x_0}{K}\right)\right\| = 0$, for all $x_0 \in \mathbb{R}^n$.

So let $\varepsilon > 0$ be given. Then for each $1 \leq j \leq n$, there exists $t_j \geq t_0$ such that $\|x(t; t_0, e_j)\| < \varepsilon$ on $[t_j, +\infty)$, since $\lim_{t\to\infty}\|x(t; t_0, e_j)\| = 0$.

Let $T = \max_{1 \leq j \leq n}\{t_j\}$. Then for any $x_0 \in \mathbb{R}^n$, with $\|x_0\| \leq 1$, and any $t \geq T$, we have

$$\|x(t; t_0, x_0)\| \leq \sum_{j=1}^n \left\|x\left(t; t_0, x_{0_j}e_j\right)\right\| = \sum_{j=1}^n |x_{0_j}| \cdot \|x(t; t_0, e_j)\|$$

$$< \sum_{j=1}^n 1 \cdot \varepsilon = n\varepsilon.$$

Then, $\|X(t)x_0\| = \|x(t; t_0, x_0)\| \leq n\varepsilon$, for all $t \geq T$ and all $\|x_0\| \leq 1$. since $x(t; t_0, x_0) = X(t)$. So, $\|X(t)\| \leq n\varepsilon$, for all $t \geq T$. Therefore, $\|X(t)\| \to 0$ as $t \to \infty$.

(5) Assume first that $\left\| X\left(t\right)X^{-1}\left(s\right)\right\| \leq Ke^{-\alpha(t-s)}$ for $t_0 \leq s \leq t$. Then by part (2) of *this theorem*, (7.2) is uniformly stable on $[t_0, \infty)$. Next, given $\varepsilon > 0$, let $\varepsilon_0 = \min\left\{\varepsilon, \frac{1}{2}K\right\}$ (we use $\frac{1}{2}K$ so that $\ln\left(\frac{\varepsilon_0}{K}\right) < 0$). Let $T\left(\varepsilon\right) = -\alpha^{-1}\ln\left(\frac{\varepsilon_0}{K}\right)$. Then, for any $t_1 \geq t_0$ and any $x_1 \in \mathbb{R}^n$ with $\|x_1\| \leq 1$, we have that for $t \geq t_1 + T\left(\varepsilon\right)$,

$$\left\| x\left(t; t_1, x_1\right)\right\| = \left\| X(t)X^{-1}\left(t_1\right)x_1\right\| \leq \left\| X(t)X^{-1}\left(t_1\right)\right\| \leq Ke^{-\alpha(t-t_1)}.$$

Note: $t - t_1 \geq T(\varepsilon)$ implies $Ke^{-\alpha(t-t_1)} \leq Ke^{-\alpha T(\varepsilon)} = \varepsilon_0 \leq \varepsilon$. Hence, $\left\| x\left(t; t_1, x_1\right)\right\| \leq \varepsilon$. Therefore, (7.2) is uniformly asymptotically stable on $[t_0, \infty)$.

Conversely, assume that $x' = A(t)x$ is uniformly asymptotically stable on $[t_0, \infty)$. Then, there exists $\delta_0 > 0$ such that, for each $0 < \varepsilon < \delta_0$, there exists $T\left(\varepsilon\right)$ such that $\|x_0\| < \delta_0$ and $t \geq t_1 + T\left(\varepsilon\right)$, then $\|x\left(t; t_1, x_0\right)\| < \varepsilon$. So, $\left\| X\left(t\right)X^{-1}\left(t_1\right)x_0\right\| < \varepsilon$, for $t \geq t_1 + T\left(\varepsilon\right)$ and $\|x_0\| < \delta_0$. Thus, $\left\| X(t)X^{-1}\left(t_1\right)\frac{x_0}{\delta_0}\right\| < \frac{\varepsilon}{\delta_0}$. But $\frac{x_0}{\delta_0}$ is an arbitrary vector with norm < 1. Consequently

$$\left\| X\left(t\right)X^{-1}\left(t_1\right)\right\| \leq \frac{\varepsilon}{\delta_0} \equiv \theta < 1, \text{ for } t \geq t_1 + T\left(\varepsilon\right). \tag{7.3}$$

Now $x' = A(t)x$ is uniformly stable on $[t_0, \infty)$, and so by part (2) of *this theorem*, $\left\| X\left(t\right)X^{-1}\left(s\right)\right\| \leq K$ for $t_0 \leq s \leq t$. So now let $t \geq t_1 \geq t_0$ and let the integer $m \geq 0$ be such that $t_1 + mT\left(\varepsilon\right) \leq t < t_1 + (m+1)T\left(\varepsilon\right)$. Then,

$$X\left(t\right)X^{-1}\left(t_1\right)$$
$$= X(t)X^{-1}\left(t_1 + mT\left(\varepsilon\right)\right) \prod_{j=m}^{1} \left[X\left(t_1 + jT\left(\varepsilon\right)\right)X^{-1}\left(t_1 + (j-1)T\left(\varepsilon\right)\right)\right].$$

So, from (7.3),

$$\left\| X(t)X^{-1}\left(t_1\right)\right\| \leq K\theta^m = \left(\theta^{-1}K\right)\left(\theta^{m+1}\right) = \left(\theta^{-1}K\right)e^{-(m+1)\alpha T(\varepsilon)},$$

where $\alpha = -T\left(\varepsilon\right)^{-1}\ln\theta$ (note $\theta < 1$ yields $\ln\theta < 0$).
Hence, $\|X(t)X^{-1}(t_1)\| \leq \theta^{-1}Ke^{-\alpha(t-t_1)}$, since $t - t_1 < (m+1)\alpha T(\varepsilon)$.
$\square$

Exercise 39. Consider the autonomous system $x' = f(x)$.

(i) Show that the stability of a constant solution $x(t; t_0, x_0) = x_0$ on $[t_0, \infty)$ implies the uniform stability of that same solution on $[t_0, \infty)$. (Hint: Show if $x(t)$ is a solution, so is $x(t+s)$ for any fixed s. If $x(t; t_0, x_0)$ is stable, then any translate can be shown to be stable.)

Note: (ii) You can similarly show that if $x(t; t_0, x_0)$ is asymptotically stable on $[t_0, \infty)$, then it is also uniformly asymptotically stable on $[t_0, \infty)$.

Definition 7.6. Consider the autonomous system $x' = Ax$. λ is an *eigenvalue of A of simple type* in case it appears in the diagonal block J_0 of the Jordan form of A.

Theorem 7.2. *The autonomous linear system $x' = Ax$ is stable iff all eigenvalues of A have nonpositive real parts and those with zero real part are of simple type. The system is strongly stable iff all eigenvalues of A have zero real part and are of simple type. The system is asymptotically stable iff all eigenvalue of A have negative real parts.*

Sketch of proof. There exists a nonsingular C such that

$$C^{-1}AC = J = \begin{bmatrix} J_0 & & & 0 \\ & J_1 & & \\ & & \ddots & \\ 0 & & & J_s \end{bmatrix},$$

where

$$J_0 = \begin{bmatrix} \lambda_1 & & & 0 \\ & \lambda_2 & & \\ & & \ddots & \\ 0 & & & \lambda_q \end{bmatrix}, \quad J_i = \begin{bmatrix} \lambda_{q+i} & 1 & 0 & \cdots & 0 \\ & \lambda_{q+i} & 1 & \cdots & 0 \\ & & \ddots & \ddots & \vdots \\ & & & \ddots & 1 \\ 0 & & & & \lambda_{q+i} \end{bmatrix}.$$

Then

$$e^{tJ} = \begin{bmatrix} e^{tJ_0} & & 0 \\ & \ddots & \\ 0 & & e^{tJ_s} \end{bmatrix}, \quad e^{tJ_o} = \begin{bmatrix} e^{t\lambda_1} & & 0 \\ & \ddots & \\ 0 & & e^{t\lambda_q} \end{bmatrix},$$

and

$$e^{tJ_i} = \begin{bmatrix} e^{\lambda_{q+i}} & te^{t\lambda_{q+i}} & \cdots & \frac{t^{m-1}}{(m-1)!}e^{t\lambda_{q+i}} \\ & \ddots & \ddots & \vdots \\ & & \ddots & te^{t\lambda_{q+i}} \\ 0 & & & e^{t\lambda_{q+i}} \end{bmatrix},$$

Then $X(t) = e^{tA} = Ce^{tJ}C^{-1}$.

Entries in e^{tA} are of the form $\sum_{j=1}^{q+s} p_j(t)e^{\lambda_j t}$, where $p_j(t)$ are polynomials in t. Hence, in the presence of stability, since $\|X(t)\|$ is bounded, all entries in $X(t)$ must be bounded. We can apply our previous theory to obtain the first statement and this will be clearly in the "iff" sense.

For the other parts, consider $X(t) = e^{tA}$ and $X^{-1}(t) = e^{-tA}$. Eigenvalues of A and $-A$ are negatives of each other. This will give eigenvalues of the zero real-parts. $\square$

Stability of the system $x' = A(t)x$ can also be studied using Floquet Theory, if the system is a periodic system. So assume the system is ω-periodic; i.e., $A(t + \omega) = A(t)$. Then by the Floquet Theory,

$$X(t) = Y(t)e^{tR},$$

where $X(t)$ is the solution of $X = A(t)X$, $X(0) = I$, and $Y(t) \in C^{(1)}(\mathbb{R})$ is nonsingular, and ω-periodic. Hence there exists $M > 0$ such that $\|Y(t)\| \leq M$, $\|Y^{-1}(t)\| \leq M$ on $\mathbb{R}$.

Recall also that $\omega R = \ln X(\omega)$ and that the eigenvalues of R are called *the characteristic exponents* of the system $x' = A(t)x$.

We now consider some special pairs of inequalities.

First,

$$\begin{cases} \|X(t)\| \leq M\|e^{tR}\|; \\ \|e^{tR}\| \leq M\|X(t)\| \quad (\text{since } e^{tR} = \overline{Y}^{-1}(t)X(t)). \end{cases}$$

Next, $X(t)X^{-1}(s) = Y(t)e^{tR}e^{-sR}Y^{-1}(s)$ and $e^{tR}e^{-sR} = Y^{-1}(t)X(t)$ $X^{-1}(s)Y(s)$ yield

$$\begin{cases} \|X(t)X^{-1}(s)\| \leq M^2\|e^{tR}e^{-sR}\|; \\ \|e^{tR}e^{-sR}\| \leq M^2\|X(t)X^{-1}(s)\|. \end{cases}$$

Also,

$$\begin{cases} \|X^{-1}(t)\| \leq \|e^{-tR}\|\|Y^{-1}(t)\| \leq M\|e^{-tR}\|; \\ \|e^{-tR}\| \leq M\|X^{-1}(t)\|. \end{cases}$$

Recall that e^{tR} is the fundamental matrix solution of

$$\begin{cases} \Phi = R\Phi, \\ \Phi(0) = I. \end{cases}$$

Via the use of the above pairs of inequalities, when compared with Theorems 7.1 and 7.2, we have the following theorem.

Theorem 7.3. *The periodic system $x' = A(t)x$ is stable on $[0, \infty)$ iff all characteristic exponents of R have nonpositive real parts and those with zero real parts are of simple type. The system is strongly stable iff all characteristic exponents of R have zero real parts and are of simple type. The system is asymptotically stable iff all characteristic exponents of R have negative real parts.*

Proof. The proof is very similar to that of Theorem 7.2. $\qquad\square$

Our next few remarks concern the stability of solutions of $X' = A(t)X$ in terms of the coefficient matrix $A(t)$.

Suppose $A(t)$ is a continuous $n \times n$ matrix functions on $[t_0, \infty)$. Let $X(t)$ be the solution of

$$\begin{cases} X' = A(t)X, \\ X(t_0) = I. \end{cases}$$

Recall that

$$\det X(t) = \det X(t_0)\, e^{\int_{t_0}^{t} \mathrm{Tr}A(s)\, ds} = e^{\int_{t_0}^{t} \mathrm{Tr}A(s)\, ds}.$$

Lemma 7.1. *If $\overline{\lim}_{t\to\infty}\mathrm{Re}\int_{t_0}^{t} \mathrm{Tr}A(s)\, ds = +\infty$, then the system $x' = A(t)x$ is unstable on $[t_0, \infty)$.*

Proof. Suppose on the contrary that the system is stable on $[t_0, \infty)$. Then there exists $K > 0$ such that $\|X(t)\| \leq K$ on $[t_0, \infty)$. This implies that all entries in $X(t)$ are bounded, hence $|\det X(t)|$ is bounded on $[t_0, \infty)$. However,

$$\varlimsup_{t\to\infty} |\det X(t)| = \varlimsup_{t\to\infty} e^{\mathrm{Re}\int_{t_0}^{t} \mathrm{Tr}A(s)\, ds} = e^{\varlimsup_{t\to\infty}\mathrm{Re}\int_{t_0}^{t} \mathrm{Tr}A(s)\, ds} \to +\infty,$$

which is a contradiction. Hence, $x' = A(t)x$ is unstable. $\qquad\square$

Lemma 7.2. *Let $x' = A(t)x$ be a stable system on $[t_0, \infty)$. Then the system is strongly stable on $[t_0, \infty)$ iff $\varliminf_{t\to\infty}\mathrm{Re}\int_{t_0}^{t} \mathrm{Tr}A(s)\, ds > -\infty$.*

Proof. Assume first that the system is strongly stable on $[t_0, \infty)$. Then, there exists $K > 0$ such that $\|X(t)\| \leq K$ and $\|X^{-1}(t)\| \leq K$ on $[t_0, \infty)$. Hence, the entries in $X^{-1}(t)$ are bounded on $[t_0, \infty)$. So, $|\det X^{-1}(t)|$ is bounded on $[t_0, \infty)$ and we know

$$\left|\det X^{-1}(t)\right| = |\det X(t)|^{-1} = e^{-\mathrm{Re}\int_{t_0}^{t} \mathrm{Tr}A(s)\, ds}.$$

So, $e^{-\operatorname{Re}\int_{t_0}^{t}\operatorname{Tr}A(s)\,ds}$ is bounded on $[t_0,\infty)$. Therefore, by an argument as in Lemma 7.1, $\overline{\lim}_{t\to\infty}-\operatorname{Re}\int_{t_0}^{t}\operatorname{Tr}A(s)\,ds < \infty$. Hence,

$$-\lim_{t\to\infty}\operatorname{Re}\int_{t_0}^{t}\operatorname{Tr}A(s)\,ds < \infty, \text{ i.e., } \lim_{t\to\infty}\operatorname{Re}\int_{t_0}^{t}\operatorname{Tr}A(s)\,ds > -\infty.$$

For the other part, assume the system is stable and that

$$\lim_{t\to\infty}\operatorname{Re}\int_{t_0}^{t}\operatorname{Tr}A(s)\,ds > -\infty.$$

Then simply retrace the steps above to conclude that $[\det X(t)]^{-1}$ is bounded on $[t_0,\infty)$. Also by the assumed stability of the system, $\|X(t)\|$ is bounded on $[t_0,\infty)$. So, the entries in $X(t)$ are bounded on $[t_0,\infty)$, hence the entries in $X^{-1}(t)$ are bounded on $[t_0,\infty)$. Hence, $\|X^{-1}(t)\|$ is bounded, thus the system is strongly stable. $\qquad\square$

Example 7.6.

(1) The statement of Lemma 7.1 is *not* "*iff*".

Consider $x'' - x = 0$. The corresponding system is

$$\begin{bmatrix} x \\ x' \end{bmatrix}' = \begin{bmatrix} 0 & 1 \\ 1 & 0 \end{bmatrix}\begin{bmatrix} x \\ x' \end{bmatrix}.$$

Solutions of $x'' - x = 0$ are $\sinh t$ and $\cosh t$ which are unbounded on $[0,\infty)$. So $x'' - x = 0$ is unstable. But $\operatorname{Tr}A(t) = 0$ yields $\overline{\lim}_{t\to\infty}\operatorname{Re}\int_0^t\operatorname{Tr}A(s)\,ds = 0$. Hence the system is unstable, but $\overline{\lim}_{t\to\infty}\operatorname{Re}\int_0^t\operatorname{Tr}A(s)\,ds < \infty$.

(2) For equations of the form $x'' + a(t)x = 0$, it follows from Lemma 7.2 that if the equation is stable on $[t_0,\infty)$, then it is strongly stable. Here

$$A(t) = \begin{bmatrix} 0 & 1 \\ -a(t) & 0 \end{bmatrix},$$

hence $\int_{t_0}^{t}\operatorname{Tr}A(s)\,ds = 0 > -\infty$. Thus the lemma is satisfied, so we have strong stability of the system, provided the system was stable to begin with.

Chapter 8

Perturbed Systems and More on Existence of Periodic Solutions

8.1 Perturbed Linear Systems

Consider $x' = A(t)x + f(t, x)$, where $A(t)$ is an $n \times n$ matrix-valued function. This equation can be considered as a perturbation of the linear system $x' = A(t)x$.

For
$$x' = f(t, x), \tag{8.1}$$
let $x(t; t_0, x_0)$ be a solution on $[t_0, \infty)$ with $x(t_0) = x_0$. For shorthand, let $x(t) \equiv x(t; t_0, x_0)$.

If $y(t)$ is also a solution of $x' = f(t, x)$, then denote $z(t) \equiv y(t) - x(t)$. Then $z(t)$ is a solution of $z' = y' - x' = f(t, y) - f(t, x) = f(t, z + x) - f(t, x)$; i.e.,
$$z' = f(t, z + x) - f(t, x). \tag{8.2}$$
If $x(t)$ is a solution of (8.1), then $y(t) = z(t) + x$ is also a solution of (8.1) iff $z(t)$ is a solution of (8.2). Moreover, $|z(t) - 0| = |y(t) - x(t)|$, and so, studying *stability of solutions of* (8.1) is the *same as* studying the *stability of the zero solution of* (8.2); i.e., a solution $x(t)$ of (8.1) has a type of stability iff the zero solution of (8.2) has the same type of stability on $[t_0, \infty)$.

Let us examine the solution $z(t)$ of (8.2) more closely. Consider formally $\frac{d}{ds} f(t, sz + x(t)) = f_x(t, sz + x(t))z$. Then
$$\int_0^1 f_x(t, sz + x(t))z \, ds = f(t, z + x(t)) - f(t, x(t)).$$
Comparing this to equation (8.2), we have $z'(t) = \int_0^1 f_x(t, sz(t) + x(t))z \, ds$. Adding and subtracting the term $f_x(t, x(t))$, we have
$$z'(t) = f_x(t, x(t))z(t) + \int_0^1 [f_x(t, sz(t) + x(t)) - f_x(t, x(t))]z(t) \, ds.$$

149

Thus $z(t)$ is a solution of the perturbed linear system $z' = A(t)z + h(t, z)$, where $A(t) = f_x(t, x(t))$ and

$$h(t, z) = \int_0^1 [f_x(t, sz(t) + x(t)) - f_x(t, x(t))]z(t)ds.$$

Hence, we look at the stability of the zero solution of this reduced system

$$z' = A(t)z + h(t, z), \tag{8.3}$$

in evaluating the stability of a solution $x(t)$ of (8.1).

In order to make such considerations, we need to assume $x(t)$ is a solution of (8.1) on $[t_0, \infty)$, and $f(t, x)$ is continuous and has continuous first partials wrt the components of x on the tube $U = \{(t, x) \mid t \geq t_0, \|x - x(t)\| < r\}$, where $r > 0$.

With these assumptions, we have

$$\frac{\|h(t, z)\|}{\|z\|} = \frac{\|\int_0^1 [f_x(t, sz + x(t)) - f_x(t, x(t))]z\, ds\|}{\|z\|}$$

$$\leq \int_0^1 \|f_x(t, sz + x(t)) - f_x(t, x(t))\|\, ds \to 0, \text{ as } \|z\| \to 0,$$

on each compact subinterval of $[t_0, \infty)$, since f_x is uniformly continuous there, i.e., $\lim_{\|z\| \to 0} \frac{\|h(t,z)\|}{\|z\|} = 0$ uniformly wrt t on each compact subinterval of $[t_0, \infty)$.

Now, in the autonomous case, if $c \in \mathbb{C}^n$ is such that $f(c) = 0$, then $x(t) \equiv c$ is a solution of $x' = f(x)$. In this case, the perturbed system (8.3) becomes $z' = f_x(c)z + h(z)$, where $f_x(c)$ is a constant matrix and $h(z) = \int_0^1 [f_x(sz + c) - f_x(c)]z\, ds$. Hence, $\lim_{\|z\| \to 0} \frac{\|h(z)\|}{\|z\|} = 0$ uniformly for all t (since no t appears).

For yet another observation, if $f(t, x)$ has period ω in t; i.e., $f(t+\omega, x) = f(t, x)$, and if $x(t)$ is a periodic solution of $x' = f(t, x)$ with period ω on the real line, and if $f(t, x)$ is continuous and has continuous first partials wrt the components of x on the tube U, the equation (8.3) becomes $z' = A(t)z + h(t, z)$, where $A(t) = f_x(t, x(t))$ is periodic with periodic ω, and $h(t + \omega, z) = h(t, z)$. From the formula and periodicity of $h(t, z)$, we have $\lim_{\|z\| \to 0} \frac{\|h(t,z)\|}{\|z\|} = 0$ uniformly on each compact subinterval of $(-\infty, \infty)$; because of the periodicity, we need consider the limit only on $[0, \omega]$ which is compact. Thus $\lim_{\|z\| \to 0} \frac{\|h(t,z)\|}{\|z\|} = 0$ uniformly on $(-\infty, \infty)$.

Theorem 8.1. *Assume that in the equation*

$$x' = A(t)x, \tag{8.4}$$

$A(t)$ *is a continuous* $n \times n$ *matrix function on* $[t_0, +\infty)$, *and assume that* (8.4) *is uniformly stable on* $[t_0, +\infty)$. *In the perturbed system*

$$x' = A(t)x + f(t, x), \tag{8.5}$$

assume that $f(t, x)$ *is continuous on* $[t_0, \infty) \times B_r(0)$, *that* $f(t, 0) \equiv 0$ *on* $[t_0, \infty)$, *and that* $\|f(t, x)\| \leq \gamma(t)\|x\|$ *on* $[t_0, \infty) \times B_r(0)$, *where* $\gamma(t)$ *is integrable on* $[t_0, t_1]$, *for each* $t_1 > t_0$, *and* $\int_{t_0}^{\infty} \gamma(t)dt < \infty$. *Then there exists* $L > 1$ *such that, for each* $t_1 \geq t_0$ *and* x_0 *with* $\|x_0\| < L^{-1}r$, *the solution* $x(t; t_1, x_0)$ *of* (8.4) *exists on* $[t_1, \infty)$ *and satisfies* $\|x(t; t_1, x_0)\| \leq L\|x_0\|$ *on* $[t_1, \infty)$.

Proof. Assume $x(t; t_1, x_0)$ is a solution of (8.5) with $t_1 \geq t_0$ and $\|x_0\| < L^{-1}r$. We will find an expression for L which satisfies the conditions of the theorem.

For notation, let $x(t) \equiv x(t; t_1, x_0)$, and let $[t_1, \omega)$ be the right maximal interval of existence of $x(t)$. Thus on $[t_1, \omega)$, $x(t)$ is a solution of $x' = A(t)x + f(t, x(t))$. Now $f(t, x(t))$ is just a function of t, hence by the Variation of the Constants Formula,

$$x(t) = X(t) X^{-1}(t_1) x_0 + \int_{t_1}^{t} X(t) X^{-1}(s) f(s, x(s)) ds, \quad \forall t \in [t_1, \omega),$$

where $X(t)$ is the solution of $X' = A(t)X$, $X(t_0) = I$. Note: $X(t)X^{-1}(t_1)$ is the solution of $X' = A(t)X$, $X(t_1) = I$.

Now by the uniform stability of (8.4), there exists $K > 0$ such that $\|X(t)\| \leq K$ and $\|X(t) X^{-1}(s)\| \leq K$, $t_0 \leq s \leq t < \infty$. Hence, for each $t_1 \leq t < \omega$, $\|x(t)\| \leq K\|x_0\| + K\int_{t_1}^{t} \gamma(s) \|x(s)\|ds$. Applying the Gronwall Inequality, we have

$$\|x(t)\| \leq \|x_0\| K e^{\int_{t_1}^{t} K\gamma(s)ds} \leq \|x_0\| K e^{\int_{t_0}^{\infty} K\gamma(s)ds} = L^*\|x_0\|,$$

where $L^* = Ke^{\int_{t_0}^{\infty} K\gamma(s) ds}$. Now this puts a "lid" on $x(t)$. Since $[t_1, \omega)$ is right maximal, recall that $x(t) \to \partial D$ as $t \to \omega$, where $D = [t_1, \omega) \times B_r(0)$, but this condition puts a "lid" on that. See Figure 8.1.

Now let $h > 1$ be fixed and set $L \equiv \max\{h, L^*\}$. Assume $\|x_0\| < L^{-1}r$; in fact, let $r_0 = L\|x_0\|$, then, $\|x_0\| = L^{-1}r_0 < L^{-1}r$. Hence, $r_0 < r$. Then we have $\|x(t)\| \leq L^*\|x_0\| \leq L\|x_0\| = r_0$ on $[t_1, \omega)$.

We claim that $\omega = +\infty$: Choose $\bar{r}$ such that $r_0 < \bar{r} < r$ and assume $\omega < \infty$. Now the set $[t_1, \omega] \times B_{\bar{r}}(0)$ (compact) $\subseteq [t_0, \infty) \times B_r(0)$. Now $f(t, x)$ is continuous on $[t_0, \infty) \times B_r(0)$. By Chapter 2, the solution approaches the boundary as $t \to \omega$, hence must leave the compact set. So, there exists τ with $t_1 < \tau < \omega$ such that $(t, x(t)) \notin [t_1, \omega] \times \overline{B_{\bar{r}}(0)}$, for all $\tau < t < \omega$.

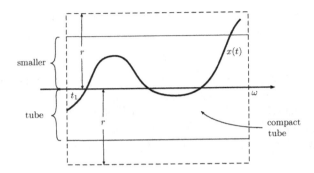

Fig. 8.1   The solution must leave each compact set (as $t \to \omega$), i.e., $x(t)$ leaves the top of the small tube, but this condition puts a lid on that.

Now $t, \tau \in [t_1, \omega)$, hence the graph leaves at the "top" of the compact set, i.e., $\|x(t)\| > \bar{r} > r_0$, for $\tau < t < \omega$. But $\|x(t)\| \leq L\|x_0\| = r_0 < \bar{r}$, for all $t_1 \leq t < \omega$, which is a contradiction. Thus, $\omega = +\infty$.

Then it follows that $\|x(t)\| \leq L\|x_0\|$ on $[t_1, \infty)$, for all $t_1 \geq t_0$.   □

Corollary 8.1. *If the hypotheses of Theorem 8.1 are satisfied, then the solution $x(t) \equiv 0$ of (8.5) is uniformly stable on $[t_0, \infty)$.*

Proof. Obviously, $x(t) \equiv 0$ is a solution since $f(t, 0) = 0$.

Given $\varepsilon > 0$ and $t_1 > t_0$, choose $\|x_0\| < \min\{\frac{1}{2}L^{-1}r, \frac{1}{2}L^{-1}\varepsilon\}$. By Theorem 8.1, $\|x(t)\| \leq L\|x_0\| < \varepsilon$. Hence, the zero solution is uniformly stable.   □

Corollary 8.2. *Assume again the hypotheses of Theorem 8.1 and in addition that the system $x' = A(t)x$ is asymptotically stable on $[t_0, \infty)$. Then the zero solution of (8.5) is both uniformly stable and asymptotically stable on $[t_0, \infty)$.*

Proof. By Corollary 8.1, the zero solution is uniformly stable.

Now let $x_0 \in \mathbb{C}^n$ and satisfy $\|x_0\| < L^{-1}r$ where L is as in Theorem 8.1. Then $x(t) = x(t; t_0, x_0)$ exists on $[t_0, \infty)$ by Theorem 8.1 and

$$x(t) = X(t)x_0 + \int_{t_0}^{t} X(t)X^{-1}(s)f(s, x(s))\, ds, \text{ on } [t_0, \infty),$$

where $X(t)$ is the solution of $X' = A(t)X$, $X(t_0) = I$. For $t \geq t_1 > t_0$, we can write

$$x(t) = X(t) x_0 + \int_{t_0}^{t_1} X(t) X^{-1}(s) f(s, x(s)) \, ds$$

$$+ \int_{t_1}^{t} X(t) X^{-1}(s) f(s, x(s)) \, ds.$$

Upon applying norm inequalities and using $\|X(t)X^{-1}(s)\| \leq K$ and $\|x(t)\| \leq L\|x_0\|$ on $t_0 \leq s \leq t < \infty$, we have

$$\|x(t)\| \leq \|X(t)\| \|x_0\| + L\|x_0\| \|X(t)\| \int_{t_0}^{t_1} \|X^{-1}(s)\| \gamma(s) ds$$

$$+ L\|x_0\| K \int_{t_1}^{t} \gamma(s) ds.$$

Given $\varepsilon > 0$, since $\int_{t_0}^{\infty} \gamma(s) \, ds < \infty$, we can choose $t_1 > t_0$ such that

$$\int_{t_1}^{t} \gamma(s) \, ds < \frac{\varepsilon}{2L\|x_0\|K}, \quad \text{for any } t > t_1.$$

From the asymptotic stability of (8.4), $\|X(t)\| \to 0$ as t $\to \infty$. Hence, we can choose $t_2 > t_1$ such that, for $t \geq t_2$, we have

$$\|X(t)\| \left\{ \|x_0\| + L\|x_0\| \int_{t_0}^{t_1} \|X^{-1}(s)\| \gamma(s) \, ds \right\} < \frac{\varepsilon}{2}.$$

Therefore, from above, for $t \geq t_2$, $\|x(t)\| < \varepsilon$. Hence, $\lim_{x \to \infty} \|x(t)\| = 0$.
Therefore, the zero solution of (8.5) is asymptotically stable. □

Corollary 8.3. *Assume that $A(t)$ and $B(t)$ are continuous $n \times n$ matrix functions on $[t_0, \infty)$ and assume that the system $x' = A(t)x$ is uniformly stable (and asymptotically stable), and assume that $\int_{t_0}^{\infty} \|B(t)\| \, dt < \infty$. Then the system*

$$x' = (A(t) + B(t)) x \tag{8.6}$$

is uniformly stable (and asymptotically stable) on $[t_0, \infty)$.

Proof. Let $f(t, x) = B(t)x$ and $\gamma(t) = \|B(t)\|$. The conclusion follows from above. □

Example 8.1. (1) The D.E.

$$x'' + x = 0 \tag{8.7}$$

is uniformly stable on $[0, +\infty)$. This is true, because all solutions are linear combinations of $\sin t$ and $\cos t$.

Consider the perturbed system

$$x'' + p(t)x' + (1 + q(t))\,x = 0. \qquad (8.8)$$

Let

$$y = \begin{bmatrix} x \\ x' \end{bmatrix},$$

then system (8.7) is

$$y' = \begin{bmatrix} 0 & 1 \\ -1 & 0 \end{bmatrix} y$$

and the perturbed system (8.8) is given by

$$y' = \begin{bmatrix} 0 & 1 \\ -1 & 0 \end{bmatrix} y + \underbrace{\begin{bmatrix} 0 & 0 \\ -q(t) & -p(t) \end{bmatrix}}_{B(t)} y.$$

Since (8.7) is uniformly stable, (8.8) is also uniformly stable on $[0, +\infty)$ if $\int_0^\infty [|q(t)| + |p(t)|]\, dt < \infty$.

(2) Let us examine the function $\gamma(s)$ in Theorem 8.1.

Now $\int_{t_0}^\infty \gamma(s)\, ds < \infty$ is a growth condition on γ as $t \to \infty$. It is also the case that "$\gamma(t) \to 0$ as $t \to \infty$ is a type of growth condition."

Is it possible that $\overline{\lim}_{t\to\infty}\gamma(t) = +\infty$ and yet $\int_{t_0}^\infty \gamma(s)\, ds < +\infty$?

Consider the function e^{-t} in Figure 8.2. Now alter this function by

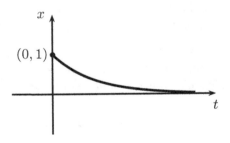

Fig. 8.2 The graph of $x = e^{-t}$.

adding spines that become more and more narrow, but taller, where say $\gamma(n) = n$, yet $\int_0^\infty \gamma(s)\, ds < \infty$. The answer to the question is in the affirmative. See Figure 8.3.

(3) In light of (2), let's consider the following example. That is, we look at an example of where we replace in Theorem 8.1, $\int_0^\infty \gamma(s)\, ds < \infty$, by $\gamma(t) \to 0$ as $t \to \infty$. The resulting problem will not be stable.

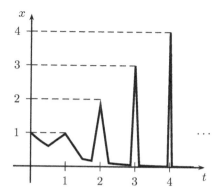

Fig. 8.3   The alternation of $x = e^{-t}$.

Let $h(t)$ be a differentiable function on $[1, +\infty)$, and define $\varphi(t) \equiv e^{\int_1^t h(s)\cos(s)\,ds}$ and define $x(t) \equiv \varphi(t)\cos t$.

Now $x(t)$ is a solution of $x'' + (1 + b(t))x = 0$, where $b(t) = 3h(t)\sin t - h'(t)\cos t - h^2(t)\cos^2 t$.

In particular, let's take $h(t) = \frac{a\cos t}{2t}$, where $a > 0$ is constant. Then

$$
\begin{aligned}
b(t) &= \frac{3a\cos t\sin t}{2t} + \frac{a\sin t\,\cos t}{2t} + \frac{a\cos^2 t}{2t^2} - \frac{a^2\cos^4 t}{4t^2} \\
&= \frac{a\sin 2t}{t} + \frac{a\cos^2 t\left(2 - a\cos^2 t\right)}{4t^2} \\
&= \frac{a\sin 2t}{t} + \beta\,(t)\,,
\end{aligned}
$$

where $\beta(t)$ is the second term in the sum. So we can write the D.E. satisfied by $x(t)$ as

$$
x'' + \left(1 + \frac{a\sin 2t}{t} + \beta(t)\right)x = 0, \tag{8.9}
$$

which can be thought of as a perturbation of

$$
x'' + \left(1 + \frac{a\sin 2t}{t}\right)x = 0. \tag{8.10}
$$

Also, $|\beta(t)| \leq \frac{a(2+a)}{4t^2}$ on $[1, +\infty)$, hence $\int_1^\infty |\beta(t)|dt < \infty$.

By Theorem 8.1, if (8.10) is uniformly stable on $[1, +\infty)$, then (8.9) is also uniformly stable on $[1, +\infty)$.

On the other hand, (8.10) is a perturbation of $x'' + x = 0$ with perturbing term $\frac{a\sin 2t}{t}x$, so that in our theorem, $\gamma(t) = \frac{a|\sin 2t|}{t}$ on $[1, \infty)$. Thus

$\gamma(t) \to 0$, as $t \to \infty$. Furthermore, $x'' + x = 0$ is uniformly stable on $[1, \infty)$. Therefore, if Theorem 8.1 were true with $\int_1^\infty \gamma(t)dt < \infty$ replaced by $\gamma(t) \to 0$ as $t \to \infty$, then (8.10) would be uniformly stable on $[1, \infty)$, hence so also would be (8.9).

But, in fact, this is not the case. If we look at the solution $x(t)$ of (8.9); i.e., $x(t) = \varphi(t)\cos t = \cos t \, e^{\int_1^t \frac{a \cos^2 s}{2s} ds}$, we see that $x(t)$ is unbounded. So, (8.9) is unstable.

In conclusion, Theorem 8.1 is false if the integral condition is replaced by $\gamma(t) \to 0$, as $t \to +\infty$.

Definition 8.1. A matrix $P \in \mathcal{M}_n$ is called a *projection* in case $P^2 = P$. In this case, if $x \in \text{Range}(P)$ and $y \in \mathbb{C}^n$ is such that $Py = x$, then $Px = P^2y = Py = x$. Moreover, $\mathbb{C}^n = [(I - P)\mathbb{C}^n] \oplus [P\mathbb{C}^n]$; i.e., any $z \in \mathbb{C}^n$ can be written as $z = x + y$, where $x \in [I - P]\mathbb{C}^n$ and $y \in P\mathbb{C}^n$, and $[(I - P)\mathbb{C}^n] \cap [P\mathbb{C}^n] = \{0\}$.

Note also that $I - P$ is a projection, so $z = (I - P)z + Pz$.

Lemma 8.1. *Let $Y(t)$ be a continuous $n \times n$ matrix function on $[t_0, \infty)$ and assume $Y(t)$ is nonsingular for all $t \geq t_0$. Assume also that P is a projection such that there exists $K > 0$ with $\int_{t_0}^t \|Y(t)PY^{-1}(s)\| ds \leq K$, for all $t \geq t_0$. Then, there exists $N > 0$ such that $\|Y(t)P\| \leq Ne^{-\frac{t}{K}}$ on $[t_0, \infty)$.*

Proof. If $P = 0$, the assertion is trivial. So assume $P \neq 0$. Then $\|Y(t)P\| \neq 0$ on $[t_0, \infty)$. Define $\varphi(t) = \|Y(t)P\|^{-1}$ which is continuous on $[t_0, \infty)$. Consider

$$\left(\int_{t_0}^t \varphi(s) \, ds \right) Y(t)P = \int_{t_0}^t \varphi(s) \, Y(t)PY^{-1}(s)Y(s)P \, ds,$$

since $P^2 = P$.

From $\int_{t_0}^t \varphi(s) \, ds > 0$ for $t > t_0$, we have

$$\left(\int_{t_0}^t \varphi(s) \, ds \right) \|Y(t)P\| \leq \int_{t_0}^t \varphi(s) \, \|Y(t)PY^{-1}(s)\| \cdot \|Y(s)P\| \, ds$$

$$= \int_{t_0}^t \|Y(t) PY^{-1}(s)\| \, ds$$

$$\leq K,$$

that is,

$$\frac{1}{K} \left(\int_{t_0}^t \varphi(s) \, ds \right) \leq \|Y(t)P\|^{-1} = \varphi(t). \tag{8.11}$$

Set $\psi(t) \equiv \int_{t_0}^{t} \varphi(s)\,ds$, so $\psi'(t) = \varphi(t) \geq \frac{1}{K}\psi(t)$, that is, $\frac{\psi'(t)}{\psi(t)} \geq \frac{1}{K}$. Hence, for any $t_1 \geq t_0$, $\int_{t_1}^{t} \frac{\psi'(s)}{\psi(s)}\,ds \geq \int_{t_1}^{t} \frac{1}{K}\,ds$. Thus, $\psi(t) \geq \psi(t_1) e^{\frac{t-t_1}{K}}$. From (8.11), $\varphi(t) \geq \frac{\psi(t)}{K}$, we have

$$\frac{1}{\varphi(t)} \leq \frac{K}{\psi(t)} \leq \frac{K}{\psi(t_1)} e^{\frac{t_1-t}{K}} \quad \text{on } [t_1, \infty).$$

Therefore,

$$\|Y(t)P\| \leq \frac{K}{\psi(t_1)} e^{\frac{t_1-t}{K}}.$$

Let $M = \max_{t_0 \leq t \leq t_1} \|Y(t)P\|$. Then $\|Y(t)P\| \leq M \leq Me^{\frac{t_1}{K}} e^{-\frac{t}{K}}$ on $[t_0, t_1]$. Let $N = \max\{Me^{\frac{t_1}{K}}, \frac{K}{\psi(t_1)}e^{\frac{t_1}{K}}\}$. Then $\|Y(t)P\| \leq Ne^{-\frac{t}{K}}$ on $[t_0, \infty)$. $\qquad\square$

Note: If $P = I$ and $Y(t)$ is continuous and nonsingular on $[t_0, \infty)$ and if for some $K > 0$, $\int_{t_0}^{t} \|Y(t)Y^{-1}(s)\|\,ds \leq K$ on $[t_0, \infty)$, then $\|Y(t)\| \leq Ne^{-\frac{t}{K}}$, for some $N > 0$.

In particular, if $Y(t) = X(t)$ is the solution of

$$\begin{cases} X' = A(t)X, \\ X(t_0) = I, \end{cases}$$

and if the condition is satisfied, then the system is asymptotically stable, since $\|X(t)\| \leq Ne^{-\frac{t}{K}} \to 0$.

Theorem 8.2. *Assume that there exists $K > 0$ such that $\int_{t_0}^{t} \|X(t) X^{-1}(s)\|\,ds \leq K$, for all $t_0 \leq t < \infty$, where $X(t)$ is the solution of $X' = A(t)X$, $X(t_0) = I$. Assume that $f(t,x)$ is continuous for $(t,x) \in [t_0, \infty) \times B_r(0)$ and that $\|f(t,x)\| \leq \gamma\|x\|$, where $0 < \gamma < \frac{1}{K}$. Then the zero solution of*

$$x' = A(t)x + f(t,x) \tag{8.12}$$

is asymptotically stable on $[t_0, \infty)$.

Proof. Let $\|x_0\| < r$ and $x(t) \equiv x(t; t_0, x_0)$ be a solution of (8.12). Let $[t_0, \omega)$ be its right maximal interval of existence. Then

$$x(t) = X(t)x_0 + \int_{t_0}^{t} X(t)X^{-1}(s)f(s, x(s))\,ds.$$

It follows from $\int_{t_0}^t \|X(t)X^{-1}(s)\| \, ds \leq K$ and Lemma 8.1 that $\|X(t)\| \leq Ne^{-\frac{t}{k}}$ on $[t_0, \infty)$. Hence, $\|X(t)\| \leq M$ on $[t_0, \infty)$ for some $M > 0$. Then

$$\|x(t)\| \leq M\|x_0\| + \int_{t_0}^t \|X(t)X^{-1}(s)\| \|f(s, x(s))\| ds$$

$$\leq M\|x_0\| + \gamma K \max_{t_0 \leq s \leq t} \|x(s)\|, \text{ for any fixed } t_0 < t < \omega.$$

By continuity, there exists $t_0 \leq \tau_t \leq t$ such that $\|x(\tau_t)\| = \max_{t_0 \leq s \leq t}\|x(s)\|$. Hence

$$\|x(\tau_t)\| \leq M\|x_0\| + \gamma K \max_{t_0 \leq s \leq \tau_t} \|x(s)\| = M\|x_0\| + \gamma K\|x(\tau_t)\|.$$

So,

$$\|x(\tau_t)\| \leq \frac{M}{1 - \gamma K}\|x_0\| \quad (\text{Note} : 1 - \gamma K > 0).$$

Therefore,

$$\|x(t)\| \leq \|x(\tau_t)\| \leq \frac{M}{1 - \gamma K}\|x_0\|,$$

which is independent of t. Hence $\|x(t)\| \leq \frac{M}{1-\gamma K}\|x_0\|$ on $[t_0, \omega)$, and from this, we conclude that if $\|x_0\|$ is chosen to be any fixed number such that $\|x_0\| < \frac{1-\gamma K}{M}r$, then for the solution $x(t; t_0, x_0)$, $\omega = +\infty$. To see this, take $\|x_0\| = \frac{1-\gamma K}{M}r_0$, where $0 < r_0 < r$ and pick r_0 so that $\frac{1-\gamma K}{M}r_0 < r$. Then $\|x(t)\| \leq r_0 < r$ on $[t_0, \omega)$. Therefore, $\omega = +\infty$ as we have seen before. Thus, if $\|x_0\| < \min\{\frac{1-\gamma K}{M}r, r\}$, then $x(t; t_0, x_0)$ exists on $[t_0, \infty)$ and $\|x(t; t_0, x_0)\| \leq \frac{M}{1-\gamma K}\|x_0\|$ on $[t_0, \infty)$. This implies the solution $x(t) \equiv 0$ of (8.12) is stable.

To complete the proof, define $\mu \equiv \overline{\lim}_{t \to \infty}\|x(t)\|$ and choose θ such that $\gamma K < \theta < 1$.

Assume $\mu > 0$. Then since $\theta^{-1}\mu > \mu$, by the definition of "limsup", there exists $t_1 > t_0$ such that $\|x(t)\| < \theta^{-1}\mu$, for all $t \geq t_1$.

Let $x(t; t_0, x_0)$ be a solution of (8.12) with the initial conditions, where $\|x_0\| < \min\{\frac{1-\gamma K}{M}r, r\}$. Then with this solution,

$$\|x(t)\| < \|X(t)\|\|x_0\| + \|X(t)\|\gamma \int_{t_0}^{t_1} \|X^{-1}(s)\| \|x(s)\| \, ds + \gamma\theta^{-1}\mu K,$$

with inequality strict by choice of θ.

Letting $t \to \infty$, since $\|X(t)\| \to 0$, we have $\mu = \overline{\lim}_{t \to \infty}\|x(t)\| \leq \theta^{-1}\mu\gamma K < \mu$; (note: $\gamma K < \theta$ yields $\gamma K\theta^{-1} < 1$).

This is a contradiction. Therefore $\mu = 0$, so that $\overline{\lim}_{t \to \infty}\|x(t)\| = 0$, or $\lim_{t \to \infty}\|x(t)\| = 0$. Hence the zero solution of (8.12) is asymptotically stable. $\qquad\square$

Exercise 40. Assume that $x' = A(t)x + f(t,x)$ satisfies the hypotheses of Theorem 8.2, and assume that the function $b : [t_0, \infty) \to \mathbb{C}^n$ is continuous and $\|b(t)\| \to 0$, as $t \to +\infty$. Prove that there exist $t_1 \geq t_0$ and $\delta_0 > 0$ such that any solution $x(t; t_1, x_1)$ of $x' = A(t)x + f(t,x) + b(t)$ with $\|x_1\| < \delta_0$ exists on $[t_1, \infty)$ and $\|x(t; t_1, x_1)\| \to 0$, as $t \to +\infty$.

We now make the observation that the integral condition in Theorem 8.2 is stronger than asymptotically stability on the unperturbed system and weaker than uniformly asymptotically stability on the unperturbed system. For example, suppose $x' = A(t)x$ is uniformly asymptotically stable on $[t_0, \infty)$. Then $\|X(t)X^{-1}(s)\| \leq Ke^{-\alpha(t-s)}$, $t_0 \leq s \leq t < \infty$, where K, α are positive constants. Then

$$\int_{t_0}^t \|X(t)X^{-1}(s)\|\, ds \leq \int_{t_0}^t Ke^{-\alpha(t-s)}\, ds = \frac{K}{\alpha} e^{-\alpha(t-s)} \Big|_{s=t_0}^{s=t}$$

$$= \frac{K}{\alpha}(1 - e^{-\alpha(t-t_0)}) \leq \frac{K}{\alpha}.$$

Thus, the integral condition is weaker than uniformly asymptotically stability of $x' = A(t)x$.

To see that the condition is strictly stronger than asymptotically stability of the unperturbed system, we proceed with an example: consider $x'' + \frac{2}{t}x' + x = 0$, with L.I. solutions $\frac{\sin t}{t}$ and $\frac{\cos t}{t}$.

The first order system is

$$y' = \begin{bmatrix} 0 & 1 \\ -1 & -\dfrac{2}{t} \end{bmatrix} y.$$

Exercise 41. Show that the zero solution is both uniformly stable and asymptotically stable on $[1, \infty)$.

Now consider the equation $x'' - \frac{2}{t}x' + x = x'' + \left(\frac{2}{t} - \frac{4}{t}\right)x' + x = 0$, with L.I. solutions $\sin t - t\cos t$, $\cos t + t\sin t$.

The corresponding perturbed system is

$$y' = \begin{bmatrix} 0 & 1 \\ -1 & -\dfrac{2}{t} \end{bmatrix} y + \begin{bmatrix} 0 & 0 \\ 0 & \dfrac{4}{t} \end{bmatrix} y$$

and $\|f(t,y)\| \leq \frac{4}{t}\|y\| \leq 4\|y\|$ on $[1, \infty)$, where

$$f(t,y) = \begin{bmatrix} 0 & 0 \\ 0 & \dfrac{4}{t} \end{bmatrix}.$$

From the solutions $\sin t - t \cos t$, $\cos t + t \sin t$, clearly the zero solution of the perturbed system is not asymptotically stable, whereas in the exercise, the unperturbed system is asymptotically stable. Therefore, the integral condition in Theorem 8.2 is strictly stronger than asymptotically stability of the unperturbed system.

Theorem 8.3. *Assume that the unperturbed system* (8.1), $x' = A(t)x$, *is uniformly asymptotically stable on* $[t_0, \infty)$, *and let* $K > 0$, $\alpha > 0$ *be such that* $\|X(t) X^{-1}(s)\| \leq Ke^{-\alpha(t-s)}$, *for* $t_0 \leq s \leq t < \infty$, *where* $X(t)$ *is the solution of*

$$\begin{cases} X' = A(t)X, \\ X(t_0) = I. \end{cases}$$

Assume $f(t,x)$ *is continuous on* $[t_0, \infty) \times B_r(0)$, *for some* $r > 0$, *and satisfies* $\|f(t,x)\| \leq \gamma\|x\|$ *for constant* γ, *with* $0 < \gamma < \frac{\alpha}{K}$. *Then every solution* $x(t; t_0, x_0)$ *of* (8.2), $x' = A(t)x + f(t,x)$ *with* $\|x_0\| < \min\{K^{-1}r, r\}$ *exists on* $[t_0, \infty)$ *and satisfies*

$$\|x(t)\| \leq Ke^{-\beta(t-s)}\|x(s)\|,$$

for all $t_0 \leq s \leq t < \infty$, *where* $\beta = \alpha - \gamma K > 0$.

Proof. Let $\|x_0\| < \min\{K^{-1}r, r\}$ and let $[t_0, \omega)$ be the maximal interval of $x(t) \equiv x(t; t_0, x_0)$, a solution of (8.2). Then, for any $t_0 \leq t_1 \leq t < \omega$,

$$x(t) = X(t) X^{-1}(t_1) x(t_1) + \int_{t_1}^{t} X(t)X^{-1}(s)f(s, x(s)) \, ds.$$

We can replace s in the hypothesis by t_1. Then

$$\|x(t)\| \leq Ke^{-\alpha(t-t_1)}\|x(t_1)\| + \gamma K \int_{t_1}^{t} e^{-\alpha(t-s)}\|x(s)\| \, ds, \, t_1 \leq t < \omega.$$

Multiplying by $e^{\alpha t}$, we have

$$e^{\alpha t}\|x(t)\| \leq K e^{\alpha t_1}\|x(t_1)\| + \gamma K \int_{t_1}^{t} e^{\alpha s}\|x(s)\|ds, \quad t_1 \leq t < \omega,$$

and then by the Gronwall inequality,

$$e^{\alpha t}\|x(t)\| \leq K e^{\alpha t_1}\|x(t_1)\|e^{\gamma K(t-t_1)}, \quad t_1 \leq t < \omega.$$

Multiplying now by $e^{-\alpha t}$, we have

$$\|x(t)\| \leq K e^{-(\alpha-\gamma K)(t-t_1)}\|x(t_1)\|, \quad \text{for } t_1 \leq t < \omega.$$

Now we can choose $t_1 = t_0$ so that

$$\|x(t)\| \leq Ke^{-(\alpha - \gamma K)(t - t_0)}\|x_0\| \leq K\|x_0\| \text{ on } [t_0, \omega).$$

By our choice of $\|x_0\|$, we have $K\|x_0\| < r$, thus $\|x(t)\| < r$, on $[t_0, \omega)$. It follows as in previous results that $\omega = +\infty$. Hence from above, we have

$$\|x(t)\| \leq Ke^{-\beta(t - t_1)}\|x(t_1)\|, \text{ for all } t_0 \leq t_1 \leq t < \infty. \qquad \square$$

Corollary 8.4. *If the hypotheses of Theorem 8.3 are satisfied, then the solution $x(t) \equiv 0$ of (8.2) is uniformly asymptotically stable.*

$\boxed{\text{Exercise}}$ **42.** Prove Corollary 8.4.

Corollary 8.5. *Assume that $A(t)$, $B(t)$ are continuous $n \times n$ matrix functions on $[t_0, \infty)$ and that $x' = A(t)x$ is uniformly asymptotically stable on $[t_0, \infty)$ and that $\|B(t)\| \to 0$, as $t \to \theta$.*

Then the system $x' = (A(t) + B(t))x$ is uniformly asymptotically stable on $[t_0, \infty)$; i.e., the zero solution is uniformly asymptotically stable.

$\boxed{\text{Exercise}}$ **43.** Prove Corollary 8.5.

Hint: If $C(t)$ is a continuous $n \times n$ matrix function on $[t_0, \infty)$ and if for some $t_1 > t_0$, $x' = C(t)x$ is uniformly asymptotically stable on $[t_1, \infty)$, then $x' = C(t)x$ is uniformly asymptotically stable on $[t_0, \infty)$. Use $\|B(t)x\| \leq \|B(t)\|\|x\| \leq r\|x\|$, if t_1 is large. Then the conclusions of Theorem 8.3, Corollary 8.1, and the previous exercise can be used.

Bibliography

Birkhoff, G. and Rota, G. C. (1989). *Ordinary Differential Equations*, 4th edn. (John Wiley & Sons, New York).

Brauer, F. and Nohel, J. A. (1969). *The Qualitative Theory of Ordinary Differential Equations: An Introduction* (W. A. Benjamin, New York).

Cesari, L. (1971) *Asymptotic Behavior and Stability Problems in Ordinary Differential Equations*, 3rd edn., Ergebnisse der Mathematik und ihrer Grenzgebiete, Band 16 (Springer-Verlag, New York).

Coddington, E. A. and Levinson, N. (1955). *Theory of Ordinary Differential Equations* (McGraw-Hill, New York), p. 54.

Coppel, W. A. (1965). *Stability and Asymptotic Behavior of Differential Equations* (D. C. Heath and Co., Boston).

Hale, J. K. (1980). *Ordinary Differential Equations*, 2nd edn. (Robert E. Krieger Publishing Co., Huntington).

Hartman, P. (1964). *Ordinary Differential Equations* (John Wiley & Sons, New York), p. 41.

Hartman, P. (2002). *Ordinary Differential Equations*, Classics in Applied Mathematics, Vol. 38 (SIAM, Philadelphia).

Kelley, W. G. and Peterson, A. C. (2010). *The Theory of Differential Equations: Classical and Qualitative*, 2nd edn., Universitext (Springer, New York).

Leighton, W. (1976). *An Introduction to the Theory of Ordinary Differential Equations* (Wadsworth Publishing Company, Belmont, CA).

Markus, L. (1955). Continuous matrices and the stability of differential system, *Math. Zeitsch* **62**, pp. 310–319.

Reid, W. T. (1971). *Ordinary Differential Equations* (John Wiley & Sons, New York).

Sanchez, D. A. (1979). *Ordinary Differential Equations and Stability Theory: An Introduction* (Dover Publications, Inc., New York), Reprint of the 1968 original.

Index

A
adjoint system, 94
asymptotically stable, 136

B
basis, 80

C
Cauchy function, 99
characteristic exponents, 123
characteristic multipliers, 123
characteristic polynomial, 104
classical solution, 1
comparison theorem, 65
complete metric space, 84
continuation of a solution, 29
continuity of solutions with respect to
   parameters, 45
continuous dependence of solutions
   on initial conditions, 42, 44
Contraction Mapping Principle, 130

D
differentiation with respect to initial
   conditions, 51
differentiation with respect to
   parameters, 55
Dini derivatives, 65

E
eigenvalue, 103
eigenvector, 103

F
first variational equation, 57
Floquet's theorem, 121
fundamental matrix solution, 94

G
Gronwall inequality, 12

H
Hill's equation, 124

I
initial value problem, 2
inner product, 83

J
Jordan canonical form, 111

K
Kamke convergence theorem, 37
Kamke uniqueness theorem, 73
kinematically similar matrices, 122

L

left maximal interval of existence, 31
linear matrix systems, 88
linear systems, 77
linearly independent, 80
Lipschitz condition, 4
logarithm of a matrix, 116

M

maximal interval of existence, 31
maximal solution, 61
metric space, 84
minimal solution, 61

N

Nagumo uniqueness result, 74
norm, 3
null space, 83

P

Peano existence theorem, 18
periodic matrix, 121
Picard existence theorem, 7, 132

Picard iterates, 8
projection, 156

R

range space, 83
Riemann integrable matrix, 87
right maximal interval of existence, 31

S

series of matrices, 85
similar matrices, 111
simple type eigenvalue, 145
solution space, 80
span, 80
stable, 47, 135, 140
strongly stable, 137

U

uniformly asymptotically stable, 137
uniformly stable, 136

V

variation of constants formula, 95